Advanced Calculus

Fundamentals of Mathematics

Authored by

Carlos Polanco

Department of Mathematics, Faculty of Sciences
Universidad Nacional Autónoma de México, México

Advanced Calculus - Fundamentals of Mathematics

Author: Carlos Polanco

ISBN (Online): 978-981-14-1508-1

ISBN (Print): 978-981-14-1507-4

© 2019, Bentham eBooks imprint.

Published by Bentham Science Publishers Pte. Ltd. Singapore. All Rights Reserved.

First published in 2019.

need for a court order if at any point you breach any terms of this License Agreement. In no event will any delay or failure by Bentham Science Publishers in enforcing your compliance with this License Agreement constitute a waiver of any of its rights.

3. You acknowledge that you have read this License Agreement, and agree to be bound by its terms and conditions. To the extent that any other terms and conditions presented on any website of Bentham Science Publishers conflict with, or are inconsistent with, the terms and conditions set out in this License Agreement, you acknowledge that the terms and conditions set out in this License Agreement shall prevail.

Bentham Science Publishers Pte. Ltd.
80 Robinson Road #02-00
Singapore 068898
Singapore
Email: subscriptions@benthamscience.net

BENTHAM SCIENCE

CONTENTS

Mathematics is the science of the connection of magnitudes. Magnitude is anything that can be put equal or unequal to another thing. Two things are equal when in every assertion each may be replaced by the other.

– Hermann Günther Grassmann
1809 – 1877

FOREWORD I

In his book, Dr. Carlos Polanco elegantly describes fundamentals of Vector Calculus and its applications. The author was able to overcome the usual gap between mathematicians and users, describing the main topics related to Vector Calculus Theory in an easily comprehensible way utilising multiple useful examples and providing exercises. This book can be used as an auxiliary book for students interested in this field as well as a reference book for the seasoned Researcher.

Vladimir N. Uversky
Russian Academy of Sciences
Pushchino, Moscow region, Russia

FOREWORD II

With Advanced Calculus, Carlos Polanco introduces an experienced view on Vector Calculus that presents a valuable reference for students as well as for their teachers. This book is very well structured, nicely written and provides a comprehensive insight into the complexity of this field. Especially, the presentation of many examples and case studies will help the readers to deepen their acquired knowledge and to relate the theory to practice. It will certainly also help researchers in related fields to refresh their knowledge and to serve as a solid and clear source on Vector Calculus. Rounding up, Carlos Polanco's book should become part of many bookshelves.

Thomas Buhse
Universidad Autonoma del Estado de Morelos
Cuernavaca Morelos, Mexico

PREFACE

Vector calculus is an essential mathematical tool for physical and natural phenomena analysis; it is a very important subject for students in the Faculties of Science and Engineering. This ebook has been designed to cover two academic semesters for second-year students, as it contains the fundamentals related to Vector Calculus. Here, the reader will find a concise and clear study of this mathematical field, it provides many examples, exercises, and a case study in each chapter. The solutions to the exercises are also included at the end of the ebook. The reader will be able to understand it even without a rigorous mathematical knowledge and will be able to immediately practise the concepts. Its main purpose is to enable students to put into practise what they are learning about the subject.

This ebook on Vector Calculus differs from others in the intensive use of affine transformations (or mapping), to modify the integration region of real-valued functions or vector-valued functions.

The first part of the ebook introduces the Vector Algebra over the real coordinate space $\mathbb{R}^n$, but cases and examples are displayed on the plane and space for a better understanding of the concepts. In this part is reviewed the dot and cross product operators, the orthogonal projection, various representations and conversions of the Cartesian and Polar coordinate systems on a plane, and the Cartesian, cylindrical, and spherical coordinate systems in a space are reviewed.

The second part is dedicated to real-valued functions, level surfaces, domain, image, graphic representations, properties of limits, continuity, differentiability, gradient, and polynomial approximation as well as the identification of critical points, the use of the Hessian matrix and Bordered Hessian matrix to characterise critical points, the implicit Function Theorem, and the inverse Function Theorem. The final chapter of this part introduces the vector-valued functions, graphic representations, and the rotational and divergence operators.

The third part presents the evaluation of integrals from mapping on a plane and in space. It separately reviews integrals of real-valued functions over bounded and unbounded regions, and curve and surface integrals of vector-valued functions over vector fields. The concept of mapping is carefully introduced, emphasising how adequate mapping simplifies integrals.

The fourth part reviews the Stokes, Gauss, and Green theorems applying mapping. It also includes examples with the complete resolution of both sides of the equalities that these theorems state.

The fifth part is a survey of exterior product and it is based on the books of Hermann Grassmann, Jerrold Marsden, and David Hestenes. We greatly appreciate the recommendation of Prof. Hestenes to include this material.

The last part covers the differential forms and the exterior product focused on line integral, surface integral, the Stokes, Gauss, and Green theorems.

The author hopes the reader interested in studying the fundamentals of calculus on several variables, finds useful the material presented here and that those who start studying this field find this information motivating. The author would like to acknowledge the Faculty of Sciences at Universidad Nacional Autónoma de México for support.

CONSENT FOR PUBLICATION

Not applicable.

CONFLICT OF INTEREST

The authors confirm that this article content has no conflict of interest.

Carlos Polanco
Department of Mathematics
Faculty of Sciences
Universidad Nacional Autónoma de México
México City
México

ACKNOWLEDGEMENTS

I would like to thank all those whose recommendations made possible the publication of this ebook

List of Credits

Case Credits   **Page**

1 **Relative Motion and Vector Addition**. Case reproduced, with altered 16
   format, with permission from: [Problem on relative motion and vector
   addition] Paul E. Hand, Massachusetts Institute of Technology 77
   Massachusetts Avenue, Cambridge, MA, USA [1].

2 **Cryptography. A frequent use of Matrix Algebr**a. Case reproduced, 18
   with altered format, with permission from: José V. Becerril, Jaime
   Grabinsky, José Guzmán, División de Ciencias Básicas e Ingeniería
   Departamento de Ciencias Básicas Universidad Autónoma
   Metropolitana Azcapozalco Av. San Pablo 180 Col. Reynosa
   Tamaulipas Delegación Azcapotzalco México, D.F [2].

3 **Equispaced Level Curves**. Case reproduced, with altered format, with 40
   permission from: [Problem on gradient, directional derivative and level
   curves] Paul E. Hand, Massachusetts Institute of Technology 77
   Massachusetts Avenue, Cambridge, MA, USA [3].

4 **Marginal Demand**. Case reproduced, with altered format, from: María 43
   Victoria Sánchez de Naranjo, Sector La Hechicera, Núcleo
   universitario Pedro Rincón Gutiérrez, Edf. A Facultad de Ciencias,
   Departamento de Matemática Mérida 5101 Venezuela [4].

5 **Minimal Surface Area**. Case reproduced, with altered format, with 59
   permission from: [Problem on optimization without constraint] Paul E.
   Hand, Massachusetts Institute of Technology 77 Massachusetts
   Avenue, Cambridge, MA, USA [5].

6 **Maximum Sales Volume**. Case reproduced, with altered format, from: 61
   Universidad de Las Palmas de Gran Canaria (ULPGC) [Copyright]
   Juan de Quesada, 30 35001 Las Palmas de Gran Canaria España [6].

7 **Potential Function of a Vector Field**. Case reproduced, with altered 71
   format, with permission from: [Problem on finding a potential function
   of a vector field] Paul E. Hand, Massachusetts Institute of Technology
   77 Massachusetts Avenue, Cambridge, MA, USA [7].

8 F_a **(x, y) = (-y/r^a,x/r^a) on the plane**. Case reproduced, with altered 73
   format, with permission from: Pete L. Clark, [HANDOUT FIVE:

List of Symbols

Symbol	Description	Page
V	Vector space.	1
$\mathbb{F}$	Field.	1
$a + b$	Vector addition.	1
α	Scalar multiplication.	1
(x, y)	Cartesian coordinate system.	3
$\|x\|$	Norm on $\mathbb{R}^3$.	4
$\dfrac{a}{\|a\|}$	Unit vector.	6
$a \cdot b$	Dot product.	6
$a \times b$	Cross product.	7
$a \cdot (b \times c)$	Triple product.	8
$v_a = \dfrac{a \cdot v}{a \cdot a}\, a$	Orthogonal projection.	9
(r, θ)	Polar coordinate system.	10
(r, θ, z)	Cylindrical coordinate system.	11
(ρ, θ, ϕ)	Spherical coordinate system.	12
$p + tv$	Equation of a line on $\mathbb{R}^3$.	14
$p + sv + tw$	Equation of a plane on $\mathbb{R}^3$.	15
$f : \mathbb{R}^n \to \mathbb{R}$	Real-valued function.	21
$\mathrm{Graph}\ (f)$	Graph of a real-valued function.	22
$f \circ g$	Composition of functions.	26
$(f \circ g)^{-1}$	Inverse composition function.	26
$\lim\limits_{x \to x_0} f$	Limit of a function.	27
$\dfrac{\partial f}{\partial x}$	Partial derivative of a function.	33
∇f	Gradient of a function.	36

$aa^\dagger$	Norm on $\mathbb{G}_2$.	128
$a(bc) = (ab)c$	Associativity on $\mathbb{G}_2$.	129
$I = \sigma_1\sigma_2$	Pseudo-vector on $\mathbb{G}_2$.	130
Ia	Clockwise rotation on $\mathbb{G}_2$.	130
aI	Counter-clockwise rotation on $\mathbb{G}_2$.	130
$\mathbb{G}_3$	Geometric algebra on $\mathbb{R}^3$.	130
$[1, \sigma_i, \sigma_j, \sigma_k, \sigma_i\sigma_j\sigma_k] \in \mathbb{G}_3$	Orthonormal basis on $\mathbb{G}_3$.	130
$\sigma_1\sigma_2\sigma_3$	Trivector.	131
$a \wedge b$	Exterior product on $\mathbb{G}_3$.	132
$a \cdot b$	Inner product on $\mathbb{G}_3$.	132
ab	Geometric product on $\mathbb{G}_3$.	130
$a(b + c)$	Distributivity of the inner product on $\mathbb{G}_3$.	132
$a \wedge (b + c)$	Distributivity of the exterior product on $\mathbb{G}_3$.	133
a^{-1}	Multiplicative Inverse on $\mathbb{G}_3$.	133
$aa^\dagger$	Norm on $\mathbb{G}_3$.	134
$a(bc) = (ab)c$	Associativity on $\mathbb{G}_3$.	135
$(x - x_0) \wedge v = 0$	Equation of a line on $\mathbb{G}_2$.	135
$(x - x_0) \wedge (u \wedge v) = 0$	Equation of a plane on $\mathbb{G}_3$.	137
$0 - \text{form}$	$0 - $ differential form.	139
dw	Exterior derivative.	140
$1 - \text{form}$	$1 - $ differential form.	140
$2 - \text{form}$	$2 - $ differential form.	140
$3 - \text{form}$	$3 - $ differential form.	140
$p - \text{form}$	$p - $ differential form.	141
$\displaystyle\int_c Pdx + Qdy + Rdz$	Line integral on $\mathbb{G}_3$.	142

Part I

PRELIMINARIES

CHAPTER 1

Vector Algebra

Abstarct: This chapter defines the concept of **vectors** on the real coordinate space $\mathbb{R}^n$, and two operators: **vector addition** and **scalar multiplication**. It explains their different representation in the Cartesian, polar, cylindrical, and spherical coordinate systems. The particular group of vectors called **unit vectors** and the operator **norm** are reviewed. Two operators with multiple geometric meaning are also studied: **dot product** and **cross product** (restricted to $\mathbb{R}^3$), as well as the **orthogonal projection** of one vector onto another and the vector equations of line and plane. The aspects here reviewed will be extensively used in the following chapters.

Keywords: Cartesian coordinates, Coordinate systems, Cross product, Cylindrical coordinates, Dot product, Line equations, Norm, Orthogonal projection, Plane equations, Polar coordinates, Scalar multiplication, Spherical coordinates, Triple product, Unit vectors, Vector addition, Vector representations.

1.1. PRELIMINARIES

The related operators would be setting of the **real coordinate space** $\mathbb{R}^n$, but cases and examples will be displayed on the plane and space for a better understanding of the concepts.

1.2. VECTOR SPACE

A vector is a mathematical abstraction that symbolises a physical magnitude from a reference system, which consists of magnitude, direction, and orientation. In this book a vector is treated as an element of a **vector space** of **finite** dimension.

Definition 1.1. A **vector space** over a **field** $\mathbb{F} \in \mathbb{R}^n$ is an algebraic structure where a set of elements called **vectors** $v, u, w \in V$, and a set of elements called **scalars** $\alpha, \beta \in F$, together with two operations, **vector addition** and **scalar multiplication**, satisfy the next eight axioms:

Property 1. $u + (v+w) = (u+v) + w$

Property 2. $u + v = v + u$

Property 3. $\exists\ 0 \in V$ called the zero vector, such that $\forall v \in V, v + 0 = v$

Property 4. $\forall\ v \in V, \exists\ -v \in V$, such that $v + (-v) = 0$

Property 5. $\alpha, \beta \in F; \alpha(\beta v) = (\alpha\beta)v$

Property 6. $1v = v$

Property 7. $\alpha(u + v) = \alpha u + \alpha v$

Property 8. $(\alpha + \beta)v = \alpha v + \beta v$

1.2.1. Vector Addition

The vector addition operation $\oplus\ : V \times V \to V$ takes two vectors $v \in \mathbb{R}^n$ and $w \in \mathbb{R}^n$ and assigns a third vector expressed as $v + w \in \mathbb{R}^n$.

Example 1.1. Let two vectors v and $w \in \mathbb{R}^2$ be over the field $\mathbb{R}$, $v = (1,2)$ and $w = (3,-1)$. What is $v + w$?

Solution1.1. If $v = (v_1, v_2)$ and $w = (w_1, w_2) \Rightarrow v + w = (v_1 + w_1, v_2 + w_2)$, then $v + w = (4,1)$.

Geometrically there are two types of vectors, those that start at the origin of the coordinate system named **fixed vectors** and those whose points of application is not start at the origin of the coordinate system, named **non-fixed vectors** (Fig. **1.1**).

Note 1.1. The addition of two **fixed vectors** yields a **fixed vector**.

1.2.2. Scalar Multiplication

The scalar multiplication operation $\otimes\ : \mathbb{F} \times V \to V$ takes any vector $v \in \mathbb{R}^n$ and a scalar $\alpha \in \mathbb{R}$ and assigns a third vector $\alpha v \in \mathbb{R}^n$, *i.e.* $\alpha v = \alpha(v_1, v_2, \cdots, v_n) = (\alpha v_1, \alpha v_2 \cdots \alpha v_n)$. When the scalar α multiplies the vector v, the length of vector αv will increase or decrease. However, if $\alpha = -1$ the vector αv keeps its length but not its orientation, which will be opposite.

Example 1.2. Given that vector $v = (-3,4) \in \mathbb{R}^2$ and scalar $\alpha = 2 \in \mathbb{R}$, what is vector αv?

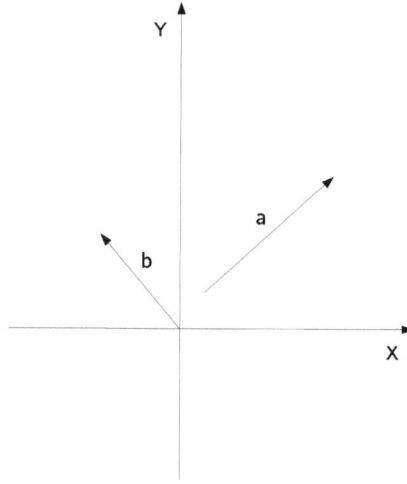

Fig. (1.1). Case in the plane. The vector a is a non-fixed vector and vector b is a fixed vector.

Solution 1.2. If $v = (v_1, v_2)$ and $\alpha \in \mathbb{R}$, then $\alpha v = \alpha(v_1, v_2) = (\alpha v_1, \alpha v_2) \Rightarrow \alpha v = 2(-3,4) = (-6,8)$.

Definition 1.2. Two vectors are the same if they have the same magnitude and direction, *i.e.* translating a vector to a new position without rotating it.

A **non-fixed vector** can be represented with the addition of **fixed vectors**. The vector c is delimited by the **fixed vectors** v and $-w$ (Fig. **1.2**), then $c = v + (-w)$.

Note 1.2. The **addition** of the vectors $v + (-w)$ is equivalent to $v - w$. This operation is known as **subtraction of vectors**.

Note 1.3. Hereafter, we will use the term **point** or **vector** indistinctly, unless otherwise specified.

1.3. CARTESIAN COORDINATE SYSTEM

Definition 1.3. A Cartesian coordinate point in $\mathbb{R}^2$ is defined as (a_1, a_2), where a_1 is the fixed perpendicular directed line on the x-axis and a_2 is the fixed perpendicular directed line on the y-axis (Fig. **1.3**). Similarly a point in the Cartesian coordinate system in $\mathbb{R}^3$ is defined as (a_1, a_2, a_3), where a_1, a_2 and a_3 are the fixed perpendicular directed lines on the axes. Both Cartesian coordinate systems can be graphically represented, however, only analytically we can represent points $(a_1, a_2, \cdots, c_n) \in \mathbb{R}^n$.

Example 1.3. What is the locus of the Cartesian coordinate point $(2,1) \in \mathbb{R}^2$?

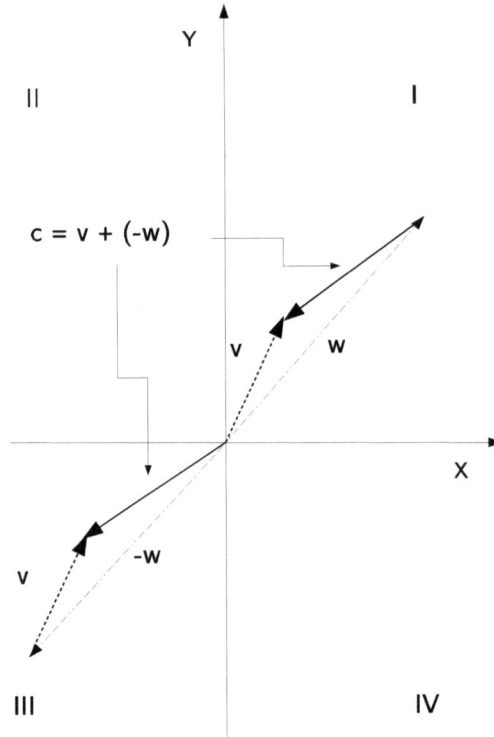

Fig. (1.2). The **non-fixed vector** $c = v + (-w)$ in Quadrant I, is the **fixed vector** c in Quadrant III.

Solution 1.3. The locus is a point located in the first Quadrant, at a distance of two units on the x-axis and a unit in the y-axis.

Definition 1.4. The **norm** (Eq. 1.1) of a **fixed vector** $a \in \mathbb{R}^n$ represents the length or distance with respect to the original point 0.

$$||a|| = \sqrt{\sum_{i=1}^{n} a_i^2}, \quad \text{where} \quad a \in \mathbb{R}^n. \tag{1.1}$$

The **norm** (Eq.1.1) of a **non-fixed vector** $c \in \mathbb{R}^n$ represents the length or **distance** (Eq. 1.2) between the **fixed vectors** $a \in \mathbb{R}^n$, and $b \in \mathbb{R}^n$.

$$||c|| = ||a - b|| = \sqrt{\sum_{i=1}^{n} (a_i - b_i)^2}, \quad \text{where} \quad c = a - b. \tag{1.2}$$

Example 1.4. There are two **fixed vectors** in a space $v = (3,0,-2)$ and $w = (1,-1,1)$. (i) What is the norm (or length) of vector v? (ii) What is the distance between the **fixed vectors** v and w?

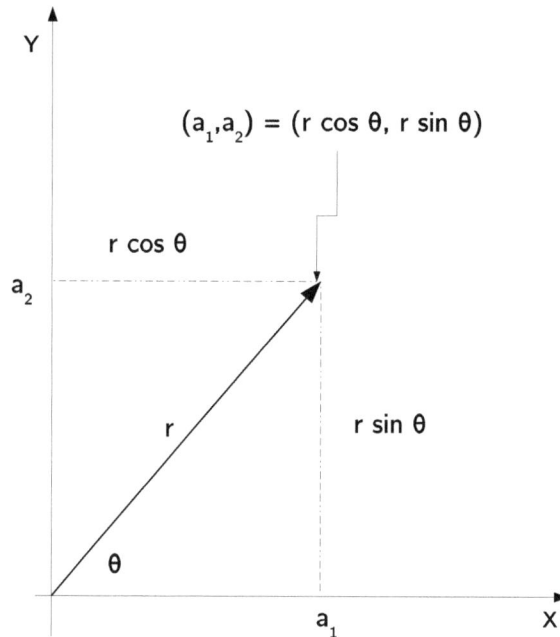

Fig. (1.3). The **Cartesian coordinate point** $(a_1, a_2) \in \mathbb{R}^2$ is a pair of numerical components, which represent their distance to the point from two fixed perpendicular directed lines, measured with the same unit of length. Note that $a_1 = r\cos\theta$ and $a_2 = r\sin\theta$.

Solution 1.4. (i) The norm of vector v is $||v|| = \sqrt{3^2 + 0^2 + (-2)^2} = \sqrt{13}$. (ii) The distance is $||v - w|| = \sqrt{(3-1)^2 + (0-(-1))^2 + ((-2)-1)^2} = \sqrt{14}$

It is important to differentiate the **norm** of a vector $||a||$ from the absolute value of a scalar $|x|$.

The **fixed vectors** in $\mathbb{R}^2$ of **norm** 1 located on the axes of the coordinate system are named **unit vectors**. Two are located on the plane $i = (1,0)$ and $j = (0,1)$; and three in $\mathbb{R}^3$: $i = (1,0,0)$, $j = (0,1,0)$, and $k = (0,0,1)$. Unit vectors can be used to represent a **fixed vector** $(a_1, a_2) = a_1 i + a_2 j = a_1(1,0) + a_2(0,1) = (a_1, 0) + (0, a_2) = (a_1, a_2)$.

Definition 1.5. Any vector $a \in \mathbb{R}^n$ can be transformed into a unit vector (Eq.1.3) by dividing it by its norm.

$$\text{The unit vector } a \text{ is } \frac{a}{||a||}. \tag{1.3}$$

Note 1.4. A Cartesian coordinate point $(a_1, a_2) \in \mathbb{R}^2$ is not completely defined if one of its components is omitted. Note that the locus of a point (a_1, α), where $\alpha \in \mathbb{R}$ represents a straight line, is parallel to the y-axis at a distance a_1 from the origin. A point (a_1, a_2) can be geometrically interpreted as the intersection of two straight lines in the Cartesian coordinate system.

1.3.1. Dot Product

Definition 1.6. The **dot product** is an algebraic operator that involves two vectors $a, b \in \mathbb{R}^n$ (Eq.1.4) and the angle θ between them (Eq.1.5).

$$a \cdot b = a_1 b_1 + a_2 b_2 + \cdots + a_n b_n \tag{1.4}$$

$$a \cdot b = ||a|| \; ||b|| \cos\theta \tag{1.5}$$

For any non-zero vectors $v, w, u \in \mathbb{R}^n$ and scalars $\alpha, \beta \in \mathbb{R}$, the **dot product** holds the next five properties:

Property 1. $v \cdot u = u \cdot v$

Property 2. $v \cdot (u + w) = (v \cdot u) + (v \cdot w)$

Property 3. $v \cdot (\alpha u + w) = \alpha(v \cdot u) + (v \cdot w)$

Property 4. $\alpha v \cdot \beta w = \alpha\beta(v \cdot w)$

Property 5. $v \perp u \Leftrightarrow v \cdot u = 0$

Note 1.5. For any two vectors $a, b \in \mathbb{R}^n$ the **dot product** is a **real number**.

Example 1.5. Given vectors $a = (1,2,3)$ and $b = (0,-2,8) \in \mathbb{R}^3$: (i) what is the dot product? (ii) what is the angle between them?

Solution 1.5. (i) $a \cdot b = (1,2,3) \cdot (0,-2,8) = 0 - 4 + 24 = 20$.

(ii) $\theta = \cos^{-1}\frac{a\cdot b}{||a||\ ||b||} = \frac{24}{||(1,2,3)||}\ ||(0,-2,8)|| = \frac{24}{\sqrt{14}\sqrt{68}} = \frac{24}{\sqrt{952}} = 0.7778 \Rightarrow$

$\theta = 0.6796$ radians.

1.3.2. Cross Product

Here, the cross product is reviewed as an operator on the plane and in space. This operator will be studied again in Geometric Calculus (Ch 9).

Definition 1.7. The **cross product** (Eq.1.6) is an algebraic operator that involves two vectors $a, b \in \mathbb{R}^3$ and the angle θ between them.

$$a \times b = \begin{vmatrix} i & j & k \\ a_1 & a_2 & a_3 \\ b_1 & b_2 & b_3 \end{vmatrix} \tag{1.6}$$

For any non-zero vectors $v, w, u \in \mathbb{R}^3$ and scalars $\alpha, \beta \in \mathbb{R}$, the **cross product** holds the next six properties:

Property 1. $v \times v = 0$

Property 2. $v \times u = -(u \times v)$

Property 3. $v \times (u + w) = (v \times u) + (v \times w)$

Property 4. $v \times (\alpha u + w) = \alpha(v \times u) + (v \times w)$

Property 5. $\alpha v \times \beta w = \alpha\beta(v \times w)$

Property 6. $||a \times b|| = ||a||\ ||b||\sin\theta$

Note 1.6. For any two vectors $a, b \in \mathbb{R}^3$ the **cross product** is a **vector**.

Note 1.7. $||a \times b||$ represents the area of a parallelogram with sides a and b.

Example 1.6. Given vectors $a = (0,2,3)$ and $b = (2,-1,-3)$. (i) What is $a \times b$? (ii) What is $||a \times b||$? (iii) What is the angle between them?

Solution 1.6.

(i) $(0,2,3) \times (2,-1,-3) = \begin{vmatrix} i & j & k \\ 0 & 2 & 3 \\ 2 & -1 & -3 \end{vmatrix} = i \begin{vmatrix} 2 & 3 \\ -1 & -3 \end{vmatrix} - j \begin{vmatrix} 0 & 3 \\ 2 & -3 \end{vmatrix} + k \begin{vmatrix} 0 & 2 \\ 2 & -1 \end{vmatrix} = (-3,6,-4)$

(ii) $||a \times b|| = ||(-3,6,-4)|| = \sqrt{(-3)^2 + 6^2 + (-4)^2} = 7.8102.$ (iii) $||a \times b|| = ||a||$

$$||b||\sin\theta \Rightarrow \theta = \sin^{-1}\frac{||a\times b||}{||a||\ ||b||} = \frac{7.8102}{\sqrt{13}\sqrt{14}} = 0.5789 \text{ radians.}$$

Note 1.8. Note that it **is not** possible to operate the **cross product** for Cartesian coordinate systems with dimension greater than 3, but it is possible with other algebras (Chapter 9), and other integrals (Chapter 10).

1.3.3. Triple Product

Definition 1.8. Geometrically, the scalar **triple product** [18] is the (signed) volume of the parallelepiped defined by the three vectors given (Fig. **1.4**).

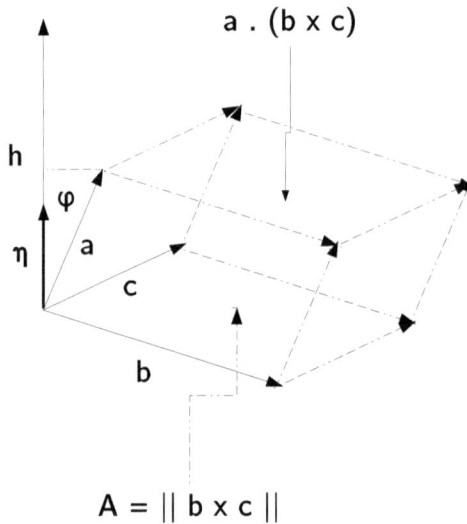

Fig. (1.4). Geometric representation of the vector named **triple product** $a \cdot (b \times c)$. Figure taken from [19].

$$a \cdot (b \times c)$$

Note 1.9. The dot product belongs to the real field $\mathbb{R}$.

Where A is the area of the parallelogram formed by the b and c vectors [19]. $a \cdot n\eta = ||a|| \, ||n||\cos\theta = ||a||\cos\phi$.   And $a \cdot (b \times c) = a \cdot An\eta = Aa\cos\phi = Ah = V$ is the volume of the parallelepiped.

1.3.4. Orthogonal Projection

Definition 1.9. The **orthogonal projection** (Eq.1.7) of vector $v \in \mathbb{R}^n$ onto vector $a \in \mathbb{R}^n$ is the vector $proj \; v_a$ (Fig. **1.5**).

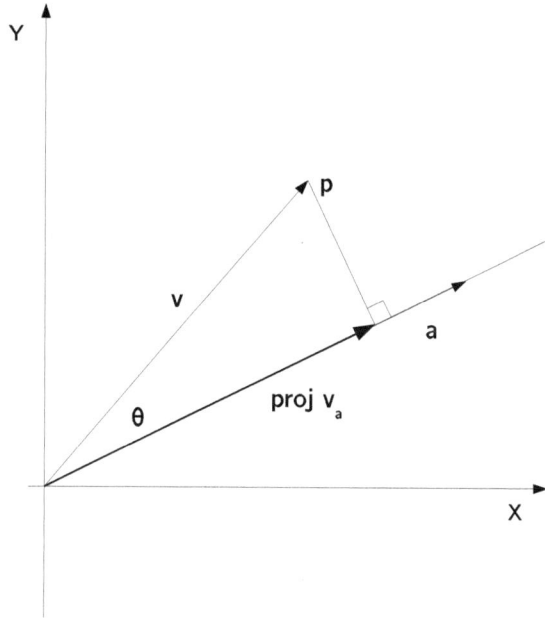

Fig. (1.5). The **orthogonal projection** $proj \; v_a \in \mathbb{R}^2$ is a segment of vector $a \in \mathbb{R}^2$, such that $p \cdot a = 0$.

$$proj \; v_a = \frac{a \cdot v}{a \cdot a} \, a \tag{1.7}$$

Example 1.7. (i) What is the orthogonal projection of $(1,2)$ onto $(1,-3)$? (ii) Is $proj \; v_a$ a vector?

Solution 1.7. (i) $proj \; v_a = \frac{a \cdot v}{a \cdot a} \, a = -\frac{1}{2}(1,-3) = (i - \frac{1}{2}, \frac{3}{2})$. (ii) Yes, it is. The component $\frac{a \cdot v}{a \cdot a}$ is a scalar.

1.4. POLAR COORDINATE SYSTEM ON THE PLANE

Definition 1.10. A polar coordinate point on a plane is represented as (r, θ), where r is the length of the fixed perpendicularly directed line that joins point $(0,0)$; and θ is the angle that forms this line with the x-axis (Fig. **1.3**).

Example 1.8. (i) Describe the locus of the polar coordinate point $(\sqrt{2}, \pi/4)$. (ii) What is its Cartesian coordinate point?

Solution 1.8. (i) It is a fixed vector of length $||(1,1)|| = \sqrt{2}$ that forms an angle $\pi/4$ radians with respect to the x-axis. (ii) The Cartesian coordinate point is $(1,1)$.

Note 1.10. A polar coordinate point (r, θ) is not sufficiently defined, if any of these parameters are omitted. Note that the point $(1, \theta)$, where θ can take any value between $[0, 2\pi]$, generates a circumference with radius 1 and centre in the origin.

1.4.1. Polar and Cartesian Coordinate Systems on the Plane

The equivalence between polar and Cartesian coordinate systems $(r, \theta) \Leftrightarrow (x, y)$, aims to identify a point on the plane $\mathbb{R}^2$ using different rules dependent on the value of the angle θ [20].

$$r = \sqrt{x^2 + y^2} \qquad \theta = 0 \qquad\qquad \text{if } x > 0, y \neq 0 \qquad (1.8)$$

$$= \tan^{-1}\frac{y}{x} + \pi \qquad \text{if } x < 0, y \geq 0$$

$$= \tan^{-1}\frac{y}{x} - \pi \qquad \text{if } x < 0, y < 0$$

$$= \frac{\pi}{2} \qquad\qquad \text{if } x = 0, y > 0$$

$$= -\frac{\pi}{2} \qquad\qquad \text{if } x = 0, y < 0$$

$$= \text{undefined} \qquad \text{if } x = 0, y = 0$$

$$x = r\cos\theta \qquad y = r\sin\theta$$

1.5. CYLINDRICAL COORDINATE SYSTEM

Definition 1.11. A cylindrical coordinate point in a space $\mathbb{R}^3$ is represented by (r, θ, z), where r is the **orthogonal projection** of the point onto the xy-plane, θ is the angle that forms r with the x-axis, and z is the length of the fixed perpendicularly directed line of the z-axis (Fig. **1.6**).

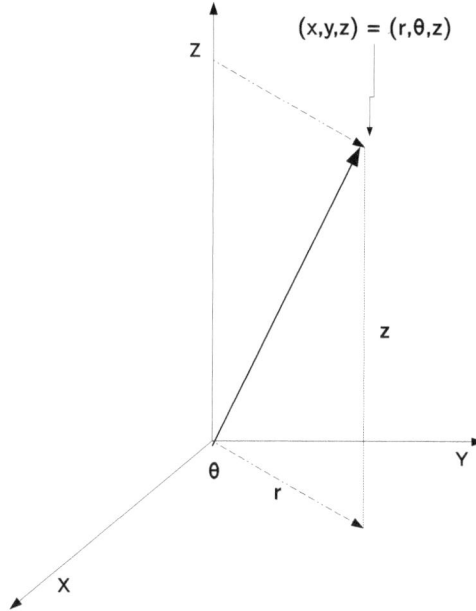

Fig. (1.6). A **cylindrical coordinate point** is the triplet (r, θ, z), where the point is projected onto the xy-plane and its length is the fixed perpendicularly directed line of the z-axis.

Example 1.9. (i) Describe the locus of the cylindrical coordinate point $(\sqrt{3}, \pi/4, 1)$. (ii) What is its Cartesian coordinate point?

Solution 1.9. (i) It is a point located in the first octant, on a line of longitude $\|a\| = \sqrt{3}$ that forms an angle $\theta = \pi/4$ radians with respect to the x-axis and onto the plane $z = 1$. (ii) The Cartesian coordinate point is $(1,1,1)$.

Note 1.11. A cylindrical coordinate point (r, θ, z) is not sufficiently defined if any of its parameters is omitted. Note that the point $(1, \theta, 1)$, where $r = 1$ θ can take any value between $[0, 2\pi]$ and $z = 1$, generates a unit circle parallel to the xy-plane at a height $z = 1$.

1.5.1 Cylindrical and Cartesian Coordinate System

The equivalence between cylindrical and Cartesian coordinate systems $(r, \theta, z) \Leftrightarrow$ (x, y, z), aims to identify a point in a space using different rules (Eqs.1.9), dependent on the value of the angle θ [21].

$$r = \sqrt{x^2 + y^2} \quad \theta = 0 \qquad \text{if } x = 0 \; y = 0 \qquad z = z \qquad\qquad \textbf{(1.9)}$$

$$= \sin^{-1}\frac{y}{r} \qquad \text{if } x \geq 0$$

$$= \tan^{-1}\frac{y}{x} \qquad \text{if } x > 0$$

$$= -\sin^{-1}\frac{y}{r} + \pi \quad \text{if } x < 0$$

$$x = r\cos\theta \qquad y = r\sin\theta \qquad\qquad z = z$$

1.6. SPHERICAL COORDINATE SYSTEM ON THE SPACE

Definition 1.12. A spherical coordinate point in a space $\mathbb{R}^3$ is described as (ρ, θ, ϕ), where ρ is the length of the line segment that joins the origin of the coordinate system with the point; θ is the angle between the vector r and the x-axis; and ϕ is the angle formed by the vector ρ and the z-axis (Fig. **1.7**).

Example 1.10. (i) Describe the locus of the spherical coordinate point $(1, \pi/4, 0)$. (ii) What is its Cartesian coordinate point?

Solution 1.10. (i) It is the positive part of the z-axis. (ii) The Cartesian coordinate point is $(0,0,1)$.

Note 1.12. A spherical coordinate point (ρ, θ, ϕ) is not sufficiently defined if any of these parameters is omitted. Note that the point $(1, \theta, \pi/4)$, where $\rho = 1$, θ can take any value between $[0, 2\pi]$ and $\phi = \pi/4$, generates a unit circle on the xy-plane.

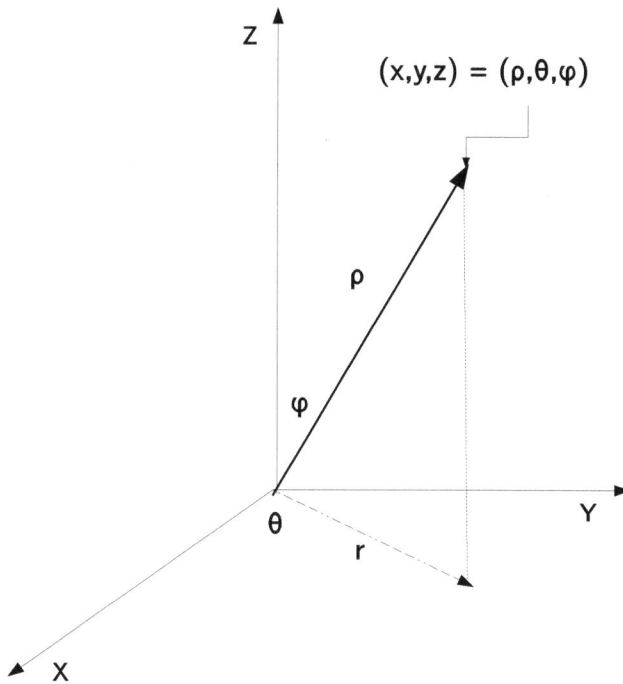

Fig. (1.7). A **spherical coordinate point** is the triplet (ρ, θ, ϕ), where the vector ρ is located in the space and limited by two angles: θ that is on the xy-plane and ϕ formed with the z-axis.

1.6.1. Spherical and Cartesian Coordinate Systems

The equivalence between spherical and Cartesian coordinate systems $(\rho, \theta, \phi) \Leftrightarrow (x, y, z)$, aims to identify a point in a space from two (Eqs. 1.10) rules [22].

$$\rho = \sqrt{x^2 + y^2 + z^2} \qquad \theta = \cos^{-1}\frac{z}{\rho} \qquad \varphi = \tan^{-1}\frac{y}{x} \qquad \textbf{(1.10)}$$

$$x = \rho\sin\phi\cos\theta \qquad y = \rho\sin\phi\sin\theta \qquad z = \rho\cos\phi$$

1.7. VECTOR REPRESENTATIONS

In this chapter, we have developed the vector algebra necessary to introduce now the vector representation of a line and a plane. The reader will find in these representations, the possibilities of this algebra using the vectors that characterise the line and the plane.

1.7.1. Equation of a Line on the Space

Definition 1.13. A line L in $\mathbb{R}^2$ or $\mathbb{R}^3$ is defined by two **fixed vectors** (Eq.1.11). The first, named **position vector**, places the line in a plane or a space; the other vector, named **direction vector** v, defines the inclination of the line (Fig. **1.8**).

$$L(t) = p + tv, \text{where } p, v \in \mathbb{R}^n, \text{and}, t \in \mathbb{R}. \tag{1.11}$$

Note 1.13. The selection of different scalar values $t \in \mathbb{R}$ produces different fixed vectors (or points), this succession of points generates the line L. Note that different **fixed vectors** p and v produce different but equivalent representations of the line L.

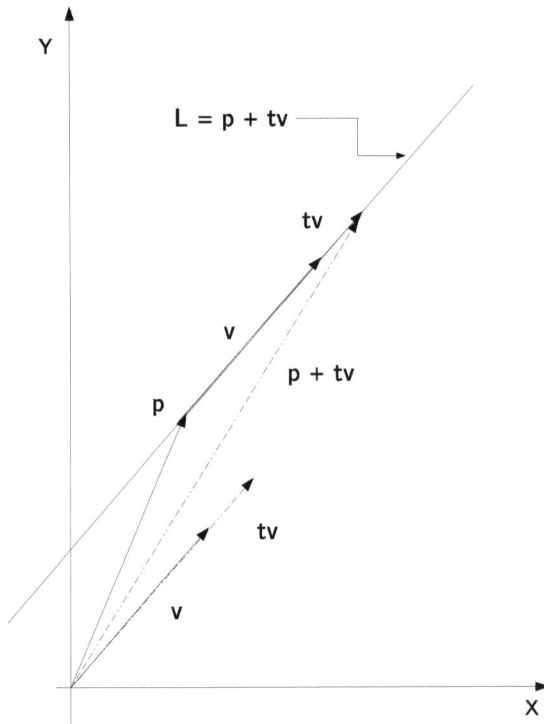

Fig. (1.8). The **line** $L \in \mathbb{R}^2$ is a succession of **fixed vectors** $p + tv, t \in \mathbb{R}$, where the position and inclination of the line is defined by the **fixed vectors** p and v respectively.

Example 1.11. Given the points $(0,0)$ and $(2,1) \in \mathbb{R}^2$, determine the line L. (ii) What is the scalar t to obtain point $(6,3)$? (iii) Is the point $(5,6)$ in line L? (iv) What is the normal line L_1 that goes through the point $(0,0)$.

Solution 1.11. (i) Given a position vector $p = (0,0)$ and a direction vector $v = (2,1)$, then the line is $L(t) = p + tv = (0,0) + t(2,1), t \in \mathbb{R}$. (ii) $(6,3) = (0,0) + t(2,1) \Leftrightarrow (6,3) = t(2,1) \Rightarrow \exists! \ t \in \mathbb{R}$, such that $6 = 2\,t$ and $3 = 1\,t$. So $t = 3$ is the solution for this system. (iii) If point $(5,6) \in$ line L, then $(5,6) = L(t) \Leftrightarrow (5,6) = (0,0) + t(2,1) \Leftrightarrow (5,6) = t(2,1)$ but ó $t \in \mathbb{R}$, therefore, the point $(5,6) \notin L$ solves the system. (iv) The normal vector w to v must meet $(w_1, w_2) \cdot (2,1) = 0 \Rightarrow 2w_1 + w_2 = 0$, if $w_2 = 1$, then $w_1 = -\frac{1}{2} \Rightarrow w = (-\frac{1}{2}, 1)$.

Proof. $(2,1) \cdot (-\frac{1}{2}, 1) = -1 + 1 = 0$. So the line that meets these requirements is $L_1 = (0,0) + (-\frac{1}{2}, 1)$.

1.7.2. Equation of a Plane

Definition 1.14. A plane $P \in \mathbb{R}^3$ is defined by three **fixed vectors** (Eq. 1.12). The **position vector** p places the plane in the coordinate system and the other two vectors, named **direction vectors** v and w, define the inclination (Fig. **1.8**).

$$P(s,t) = p + sv + tw, \text{where } p, v, \text{and}, \ w \in \mathbb{R}^n, \text{and}, \ s, t \in \mathbb{R}. \textbf{(1.12)}$$

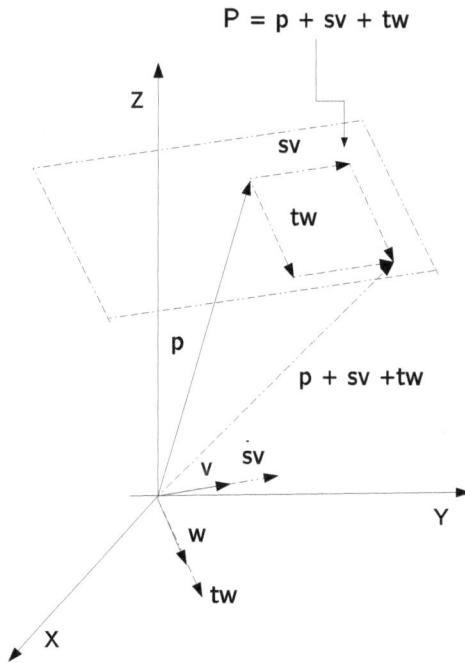

Fig. (1.9). A **plane** $P \in \mathbb{R}^3$ is a sequence of the **fixed vectors** $p + sv + tw, s, t \in \mathbb{R}$, where the plane P is a product of the linear combination $sv + tw$.

Definition 1.15. Given a set of **fixed vectors** $v_i \in V \subset \mathbb{R}^n$, vector v is a **linear combination** of V (Eq. 1.3) if $\exists$ $\alpha_1, \ldots, \alpha_n \in$ field $\mathbb{R}$, such that

$$v = \sum_{i=1}^{n} \alpha_i v_i \ \exists \ \alpha_1, \ldots, \alpha_n \in \mathbb{R} \text{ such that } v = \sum_{i=1}^{n} \alpha_i v_i \ \exists \ \alpha_1, \ldots, \alpha_n \in V. \qquad (1.13)$$

1.8. CASE STUDY: RELATIVE MOTION AND VECTOR ADDITION

Case 1.1. A river flows with a speed of $15m/s$ in a northeast direction. A particular boat can propel itself at a speed of $30\frac{m}{s}$ relative to the water. What direction must the boat point to go west? **Case adapted with permission of the author** [1].

We introduce symbols for the three speeds of this problem. Let v_{river} be the speed of the river. Let v_{boat} be the speed of the propulsion of the boat, and v_{net} be the net speed of the boat relative to the land. We now look for a relation between these speeds.

Note 1.14. When an object has a speed relative to a moving medium, its net speed is the sum of its relative speed and the speed of the medium.

Then

$$v_{net} = v_{river} + v_{boat}$$

Graphically (Fig. **1.10**), this relation is represented

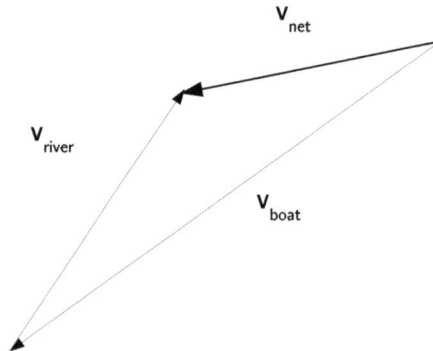

Fig. (1.10). Graphical representation of v_{net}, v_{river}, and v_{boat}.

Then, we express the problem statement in terms of vectors. Given that the river has length 15 and it is in the direction $i + j$, the j component of v_{net} is 0 and the length of v_{boat} is 30. Let us find the direction of v_{boat}. First, we find v_{river} and we characterize a general v_{boat}. Then, we select a vector such that the sum has no

j component. To find v_{river} we determine the vector from its length and direction for any nonzero vector x

$$x = ||x||\text{dir}x. \tag{1.14}$$

Note 1.15. To determine x from its length and direction, we have $x = ||x||\mathbf{dir}x$.

Knowing that v_{boat} has length 15 and it is in the direction $i + j$, we have $v_{boat} = 15\text{dir}(i + j)$. With (Eq. 1.14) we compute $\text{dir}(i + j) = \frac{i+j}{\sqrt{2}}$ to get

$$v_{boat} = 15\frac{i + j}{\sqrt{2}}.$$

Then, we characterise the possible speed range of the boat $v_{boat} = ai + bj$. Combining v_{river} and v_{boat} we get $v_{net} = \frac{15}{\sqrt{2}}(i + j) + (ai + bj)$.

Note 1.16. Vector component addition. The sum of two vectors is obtained by adding each component $\langle x_1, x_2 \rangle + \langle y_1, y_2 \rangle = \langle x_1 + y_1, x_2 + y_2 \rangle$.

Setting the **j** component to zero, we conclude $b = -\frac{15}{\sqrt{2}}$. To solve for a, we have the information that v_{boat} has length 30, then $30 = \sqrt{a^2 + b^2}$

Note 1.17. Definition of vector length: $||\mathbf{x}|| = \sqrt{x_1^2 + x_2^2}$

Hence,

$$a^2 = 787.5$$

$$a = \pm\sqrt{787.5}$$

In order to make the boat travel west, a must be negative

$$v_{boat} = -\sqrt{787.5}\ i - \frac{15}{\sqrt{2}}\ j \tag{1.15}$$

$$\approx -28.0\ i - 10.6\ j \tag{1.16}$$

The direction the boat must go is

$$\mathrm{dir}v_{boat} = \frac{-\sqrt{787.5}\ i - \dfrac{15}{\sqrt{2}}\ j}{30} \approx -0.935i - 0.353j.$$

1.9. CASE STUDY: CRYPTOGRAPHY. A FREQUENT USE OF MATRIX ALGEBRA

Case 1.2. **Definition 1.16.** A vector column **x** is in fact a matrix $m \times 1$, *i.e.* a matrix with only one column of m elements [23], where

$$\mathbf{x} = (x_1, x_2, \dots, x_m)^{\mathrm{T}} = \begin{pmatrix} x_1 \\ x_2 \\ \vdots \\ x_m \end{pmatrix}$$

Cryptography. Text encryption requires matrices and vectors. **Case adapted with permission of the author** [2].

From this definition, we will explain how to encrypt and decrypt a text. For this purpose, we will use the letters of the alphabet (Table **1.1**) and include the symbol of blank space.

Table 1. Codes used for the letters of the alphabet [24].

a	b	c	d	e	f	g	h	i	j	k	l	m	n	o	p	q	r	s	t	u	v	w	x	y	z	
81	14	27	42	127	22	20	60	69	1	07	40	24	67	75	19	0	59	63	90	27	89	23	1	19	2	99

Let us take the next sentence " **normative validity**", to be rewritten with its numeric equivalent (Table **1.1**)

$$\{67, 65, 59, 24, \mathbf{81}, 90, 69, 89, 127, 89, 81, 40, 69, 42, 69, 90, 19\}$$

Now, this set will be divided into a set of vectors β_i with length 3.

$$\begin{pmatrix} 67 \\ 75 \\ 59 \end{pmatrix}, \begin{pmatrix} 24 \\ 81 \\ 90 \end{pmatrix}, \begin{pmatrix} 69 \\ 89 \\ 127 \end{pmatrix}, \dots, \begin{pmatrix} 69 \\ 90 \\ 19 \end{pmatrix}$$

Note 1.18. Note that the length of the vector is given arbitrarily, it can be of any length. Only add blank spaces at the end of the last vector or add a symbol at the beginning and at the end of a sentence.

Now we will define an invertible matrix **A** of 3×3 dimension, that will be called " **seed**".

$$\mathbf{A} = \begin{pmatrix} 1 & -1 & 2 \\ -2 & 0 & 4 \\ 0 & -2 & 7 \end{pmatrix}$$

Note 1.19. Matrix **A** can be any invertible matrix that will be there to encrypt and decrypt.

Finally, we compute the product **A** β_i for each vector β_i. For example: $i = 1$

$$\mathbf{A}\,\beta_1 = \begin{pmatrix} 1 & -1 & 2 \\ -2 & 0 & 4 \\ 0 & -2 & 7 \end{pmatrix}\begin{pmatrix} 67 \\ 65 \\ 59 \end{pmatrix} = \begin{pmatrix} 120 \\ 102 \\ 283 \end{pmatrix}$$

Each of these vectors $\mathbf{A}\beta_i$ is a part of the encrypted sentence $\{120,102,283\}$.

In order to decrypt each vector $\mathbf{A}\beta_i$, we use the inverse matrix $\mathbf{A}^{-1}$, and compute the product $\mathbf{A}^{-1}$ with each vector $\mathbf{A}\beta_i$.

$$\mathbf{A}^{-1} = \begin{pmatrix} 4 & \frac{3}{2} & -2 \\ 7 & \frac{7}{2} & -4 \\ 2 & 1 & -1 \end{pmatrix}$$

$$\mathbf{A}^{-1}\,\mathbf{A}\,\beta_1 = [\mathbf{A}^{-1}\,\mathbf{A}]\,\beta_1 = \begin{pmatrix} 4 & \frac{3}{2} & -2 \\ 7 & \frac{7}{2} & -4 \\ 2 & 1 & -1 \end{pmatrix}\begin{pmatrix} 120 \\ 102 \\ 283 \end{pmatrix} = \begin{pmatrix} 67 \\ 65 \\ 59 \end{pmatrix}$$

The set $\{67,65,59\}$ corresponds to the original vector β_1.

1.10. EXERCISES

Exercise 1.1. Show that the equation $A(x - x_0) + B(y - y_0) + C(z - z_0) = 0$ represents a plane in the space $\mathbb{R}^3$, with the points (x_0, y_0, z_0) and that its normal vector n is (A, B, C).

Exercise 1.2. A parallelogram has the vertices $A = (-1, 2, 3), B = (2, 0, -1)$, and $D = (3, 1, 4)$. What are the possible locations for the fourth vertex C?

Exercise 1.3. Find vector w of length 14, which is parallel to the line $21x - 80y = 13$.

Exercise 1.4. Use cross products to find the area of the triangle with vertices $P = (1, 1, 0), Q = (1, 0, 1)$, and $R = (0, 1, 1)$.

Exercise 1.5. Find the area of the parallelogram spanned by the vectors $u = (1, 2, 0)$ and $v = (a, b, c)$.

Exercise 1.6. Find the distance from point $P = (1, 4, 1)$ to the plane $2x + 3y - z = -1$.

Exercise 1.7. Find the plane perpendicular to the planes $x + y - z = 1$ and $2x - 3y + 4z = 5$ that passes through point $P = (1, 0, -2)$.

Exercise 1.8. Determine if these three points $P = (1, -5, 2)$, $Q = (-1, -3, 3)$, and $R = (-3, -1, 5)$ lie on the same line.

Part II
DIFFERENTIATION

CHAPTER 2

Real–Valued Functions

Abstract: This chapter focuses on the characterisation of a **real-valued function**, its **graphs**, and **level surfaces**; its **limits**, **continuity**, and **differentiation**. The operators here reviewed are **gradient, directional derivatives**, and the polynomial approximation to a function named **Taylor´s theorem**. All of them will be extensively used in the following chapters.

Keywords: Composition, Continuity, Differentiation, Directional derivatives, Domain of function, Gradient, Graph function, Hospital's Rule, Image of function, Level surfaces, Limits, Partial derivatives, Real-valued functions, Taylor's theorem.

2.1. PRELIMINARIES

The related operators would be setting of the **real coordinate space** $\mathbb{R}^n$, but cases and examples will be displayed on the plane and space for a better understanding of the concepts.

2.2. REAL-VALUED FUNCTIONS

The **Functions of several variables** whose domain is located on $\mathbb{R}^n$, use the same concepts applied for the calculus of one variable. However, it is necessary to introduce new techniques that enable the construction of graphs, calculate the limits, continuity, differentiation, and integration. This chapter describes and illustrates in detail these techniques.

L'Hospital's Rule is a useful tool in the resolution of the limits of a function of one variable, but also to solve the limits of a function of several variables; as will be seen, the limit of a function of several variables will be the limit of a function of one variable by trying different trajectories.

Definition 2.1. A real-valued function f is a function whose domain D_f is a set U such that $U \subset \mathbb{R}^n$, and with image I_f contained in set M such that $M \subset \mathbb{R}$. A useful equivalent notation to a real-valued function is $f \colon U \subset \mathbb{R}^n \to M \subset \mathbb{R}$.

Definition 2.2. The domain D_f (Eq. 2.1) of a real-valued function $f: U \subset \mathbb{R}^n \to M \subset \mathbb{R}$ is the set U, such that the function is defined by

$$D_f = \{(x_1, x_2, \dots, x_n) \in \mathbb{R}^n \mid \exists\, z \in \mathbb{R}: f(x_1, x_2, \dots, x_n) = z\} \tag{2.1}$$

Definition 2.3. The image I_f (Eq. 2.2) of a real-valued function $f: U \subset \mathbb{R}^n \to \mathbb{R}$ is the set $M \subset \mathbb{R}$.

$$I_f = \{z \in \mathbb{R} \mid \exists\, (x_1, x_2, \dots, x_n) \in \mathbb{R}^n: f(x_1, x_2, \dots, x_n) = z\} \tag{2.2}$$

Example 2.1. Let the real-valued function $f: U \subset \mathbb{R}^n \to \mathbb{R}, x_1 + x_2, + \cdots, +x_n$ (i) What is the domain? (ii) What is the image?

Solution 2.1. (i) The domain D_f is the space $\mathbb{R}^n$. (ii) The image I_f is the space $\mathbb{R}$.

Example 2.2. Let $f(x, y) = x^2 + y^2$. (i) Describe this function using an equivalent notation. (ii) What is the domain of f? (iii) What is the image of f? (iv) Where is the domain located? (v) Where is the image located?

Solution 2.2. (i) $f: U \in \mathbb{R}^2 \to \mathbb{R}, x^2 + y^2$. (ii) The domain is $\mathbb{R}^2$. (iii) The image is $\mathbb{R}$. (iv) The domain is located in the xy-plane (v) The image is located on the z-axis.

Example 2.3. Let $f(x, y, z) = \frac{4x^2 + 6y^2 - xyz}{x - 3}$. (i) What is the domain of f? (ii) What is the image of f?

Solution 2.3. (i) The domain is $\mathbb{R}^3 - \{(3, y, z)\}$. (ii) The image is $\mathbb{R}$.

Example 2.4. Let $f(x, y) = \sqrt{xy}$. (i) Describe this function using an equivalent notation. (ii) What is the domain of f? (iii) What is the image of f?

Solution 2.4. (i) $f: \mathbb{R}^2 \to \mathbb{R}, \sqrt{xy}$. (ii) The domain is $\mathbb{R}^2$ such that $xy \geq 0$. (iii) The image is $\mathbb{R}$.

2.2.1. Graph of a Function

Definition 2.4. The graph of a function $f: U \subset \mathbb{R}^n \to \mathbb{R}$ (Eq. 2.3) is the geometric representation of set $(x_1, x_2, \dots, x_n, f(x_1, x_2, \dots, x_n)) \in \mathbb{R}^{n+1}$.

$$\text{graph } (f) = \{(x_1, x_2, \ldots, x_n, x_{n+1}) \in \mathbb{R}^{n+1} \mid (x_1, x_2, \ldots, x_n) \in U\} \tag{2.3}$$

Example 2.5. Let the real-valued function $f(x, y) = x^2 + y^2$. Draw the graph of f and describe D_f, I_f.

Solution 2.5. The graph of function f is a paraboloid of revolution oriented upward from the origin around the z-axis (Fig. **2.1**). Its domain D_f is the space $\mathbb{R}^2$ located in the xy-plane and its image I_f is the space $\mathbb{R}$ located on the z-axis.

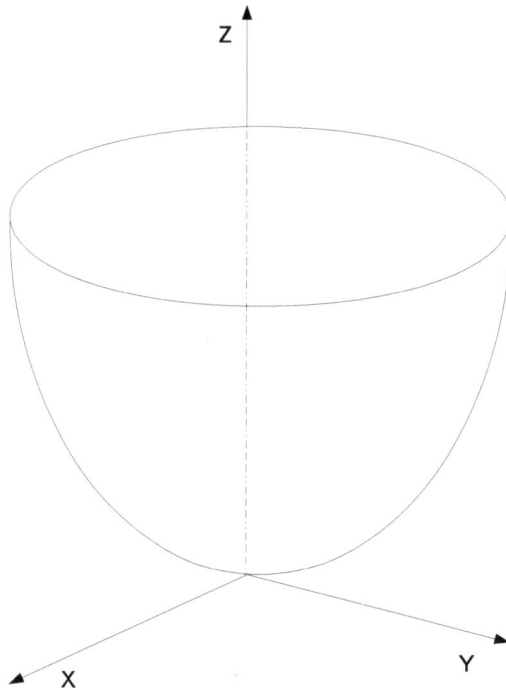

Fig. (2.1). The graph of function $f: \mathbb{R}^2 \to \mathbb{R}, x^2 + y^2$ is the set of points $(x, y, x^2 + y^2) \in \mathbb{R}^3$, its domain D_f is the space $\mathbb{R}^2$, and its image I_f is the space $\mathbb{R}$.

2.2.2. Level Surfaces

Definition 2.5. Let $f: U \subset \mathbb{R}^n \to \mathbb{R}$ and $c \in \mathbb{R}$. The level surface (Eq. 2.4) of value c is the set of points $x \in U$, where $f(x_1, x_2, \ldots, x_n) = c$. The **level surface** is always in the **domain** of function f.

The level surface $c = \{(x_1, x_2, \ldots, x_n) \in U \mid f(x_1, x_2, \ldots, x_n) = c\}$ $\qquad$ **(2.4)**

Example 2.6. Let the real-valued function $f(x,y) = x^2 + y^2$, whose graph is a paraboloid of revolution (Fig. **2.1**). What is the level surface?

Solution 2.6. The level surface is formed by the circumferences $x^2 + y^2 = c_i, i = \{0,1,2,3,\cdots\}$ in the xy-plane (Fig. **2.2**).

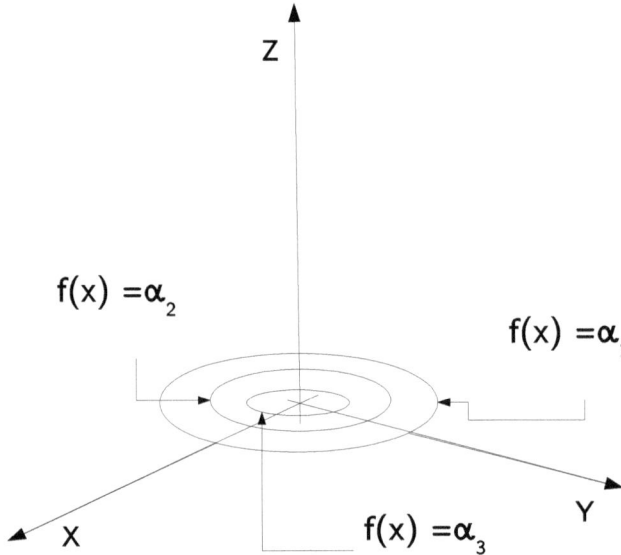

Fig. (2.2). Sample of level surface of function $f: \mathbb{R}^2 \to \mathbb{R}, x^2 + y^2$ formed by three circumferences in the xy-plane, whose radios are α_1, α_2 and α_3.

2.2.3. Construction of Graphs

It is not obvious to get a graph of a function in space $\mathbb{R}^3$, even if a specialised software is used. A graph may not be symmetrical to any of the coordinate planes, so it is not possible to know a priori how many points (or regions) in the domain of function f it is necessary to evaluate. In the case of the graph of a function that is symmetrical to one of the coordinate axes, it is possible to use (Eq. 2.5) the intercepts (or projections) of the graph on the plane.

$$
\begin{aligned}
xz - \text{plane} \cap \text{graph} f &= \{(x,y,z) \in \mathbb{R}^3 \mid f(x,0) = z \colon \mathbb{R}^2 \to \mathbb{R}\} \\
yz - \text{plane} \cap \text{graph} f &= \{(x,y,z) \in \mathbb{R}^3 \mid f(0,y) = z \colon \mathbb{R}^2 \to \mathbb{R}\} \qquad \textbf{(2.5)} \\
xy - \text{plane} \cap \text{graph} f &= \{(x,y,z) \in \mathbb{R}^3 \mid f(x,y) = z \colon \mathbb{R}^2 \to \mathbb{R}\}
\end{aligned}
$$

Example 2.7. Let the function f, $f(x,y) = x^2 + y^2$ What are the projections on the planes?

Solution 2.7.

$$xz - \text{plane} \cap \text{graph} f = \{(x,y,z) \in \mathbb{R}^3 \mid y = 0, f(x) = x^2\}$$

$$yz - \text{plane} \cap \text{graph} f = \{(x,y,z) \in \mathbb{R}^3 \mid x = 0, f(y) = y^2\}$$

$$xy - \text{plane} \cap \text{graph} f = \{(x,y,z) \in \mathbb{R}^3 \mid z = 0, x^2 + y^2 = \alpha \in \mathbb{R}\}$$

The projections of graph f on the xz plane and the yz-plane are similar parabolas (Fig. **2.3**). The projection on the xy-plane is a level surface (Fig. **2.2**).

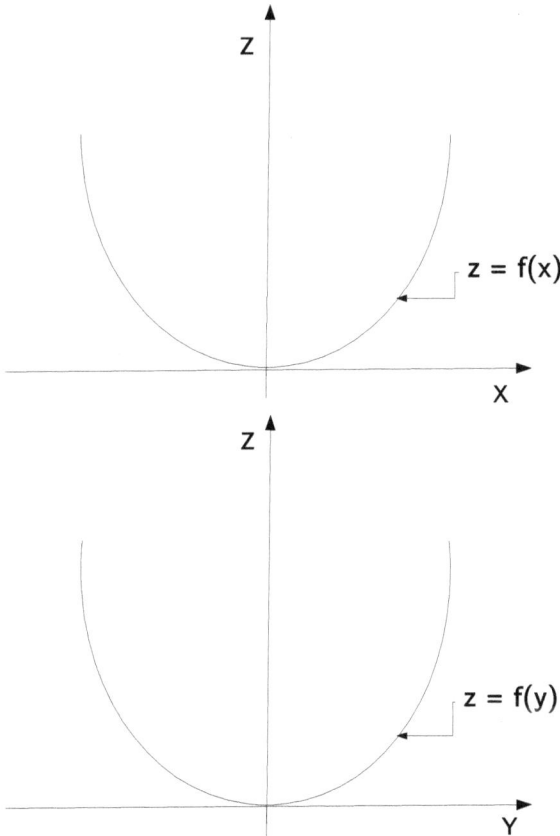

Fig. (2.3). The projections of the graph f on the xz-plane and the yz-plane are the parabolas $f(x)$ on the xz-plane and $f(y)$ on the yz-plane.

Example 2.8. Let the function f, $f(x,y) = x^2 + y^2 + z^2$. Describe its components.

Solution 2.8. The graph $f \in \mathbb{R}^4$ **cannot** be visualised directly. The domain D_f is the space $\mathbb{R}^3$. The image I_f is the space $\mathbb{R}$. The **level surface** is a set of concentric spheres.

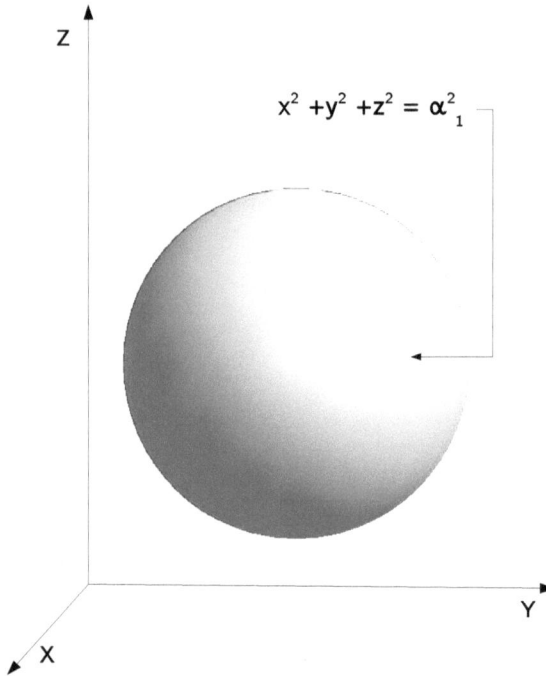

Fig. (2.4). The level surface of graph f is a set of concentric spheres. The graph of f **cannot** be visualised directly because it belongs to space $\mathbb{R}^4$.

2.3. COMPOSITION OF FUNCTIONS

In order to understand the properties of this operator, the family of the real-valued functions has been extended to consider the image of the function in $\mathbb{R}^m$, where $m > 1$.

Definition 2.6. Let the real-valued functions $f \colon \mathbb{R}^m \to \mathbb{R}^p$ and $g \colon \mathbb{R}^n \to \mathbb{R}^m$. The composition function is a **function** $f \circ g \colon f(g(x)) \colon \mathbb{R}^n \to \mathbb{R}^p$. Note that the domain of function $f \circ g$ is the domain of function g and the image of $f \circ g$ is the image of function f. The composition function $f \circ g$ is different from the functions f or g (Fig. **2.5**).

Note 2.1. The inverse composition function is $(f \circ g)^{-1} = g^{-1} \circ f^{-1}$, if there exists f^{-1} and g^{-1} (Sect. 3.5).

Example 2.9. Let $f: \mathbb{R} \to \mathbb{R}, 3x^2$ and $g: \mathbb{R} \to \mathbb{R}, \frac{x}{2}$. The composition function is the **function** $f \circ g: f(g(x)) = 3(\frac{x}{2})^2$.

Example 2.10. Let $f: \mathbb{R}^2 \to \mathbb{R}, e^{xy}$ and $g: \mathbb{R}^2 \to \mathbb{R}, x - y$. The composition function is the **function** $f \circ g: f(g(x)): \mathbb{R}^2 \to \mathbb{R}, e^{(x-y)y} = e^{(xy-y^2)}$.

Example 2.11. Let $f: \mathbb{R}^2 \to \mathbb{R}^2, (e^y, -10^x)$ and $g: \mathbb{R}^2 \to \mathbb{R}^2, (\log x, \ln y)$. The composition function is the **function** $f \circ g: f(g(x)): \mathbb{R}^2 \to \mathbb{R}^2, (y, -x)$.

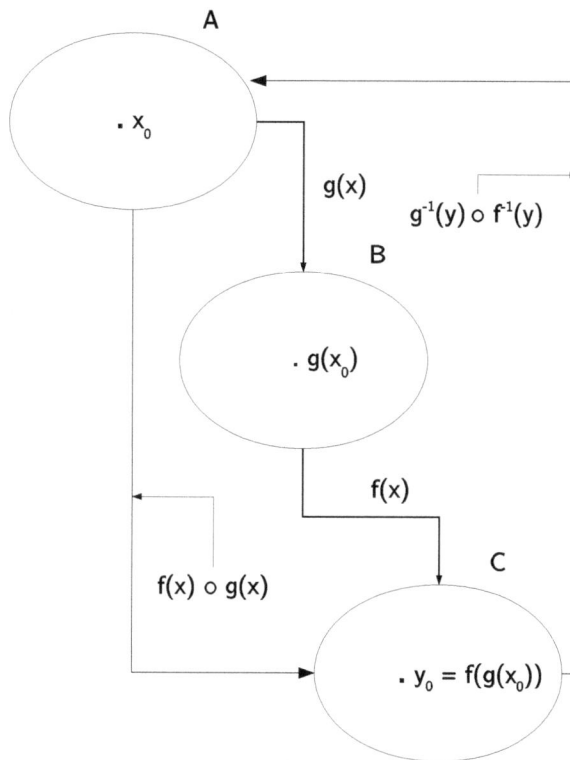

Fig. (2.5). The **function** $f(x) \circ g(x)$ goes from set A to C and the **function** $g^{-1} \circ f^{-1}$ goes from set C to A.

2.4 LIMIT OF A FUNCTION

The **limit** operator certainly is the most important mathematical operator; its concept was attributed to Augustin-Louis Cauchy in 1821. This topic is so broad, that a book can be dedicated to it. For this reason, only the most useful applications related to vector calculus will be addressed here.

Definition 2.7. Let $f: U \subset \mathbb{R}^n \to W \subset \mathbb{R}$ and let x_0 be in U or in a boundary point of U (Fig. **4**). Then $\lim_{x \to x_0} f(x) = L$ **if and only if** for every number $\varepsilon > 0$ there is a $\delta > 0$, such that for any $x \in U$ satisfying $0 < ||x - x_0|| < \delta$ we have $||f(x) - L|| < \varepsilon$ [25].

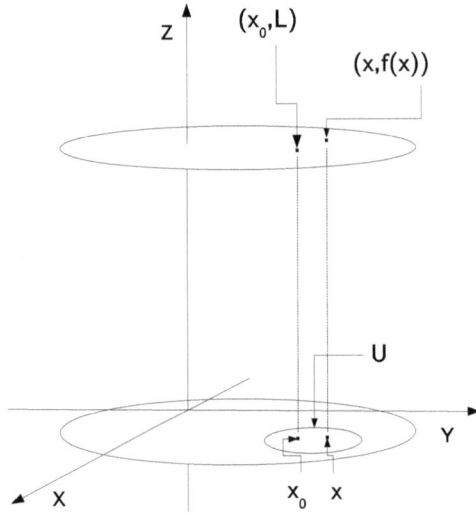

Fig. (2.6). The **limit operator** is defined in terms of the lenght $0 < ||x - x_0|| < \delta$ and $||f(x) - L|| < \varepsilon$.

Note 2.2. For the elements x and $x_0 \in U \subset \mathbb{R}^n$, the $\lim_{x \to x_0} f(x) = L$ is never equal to x_0. In fact, the function may not even be defined at x_0, yet the limit L may still exist. While x_0 may not be in the domain of f, point x when $x \to x_0$, is always in the domain of f.

Example 2.12. Show that $\lim_{(x,y) \to (0,0)} x = 0$ using the method $\varepsilon - \delta$.

Proof. $\forall \varepsilon > 0 \ \exists \ \delta > 0$ such that $|x| < \delta$ and $|y| < \delta \Rightarrow |x - 0| < \varepsilon$. But $|x - 0| < \varepsilon \Leftrightarrow |x - 0| \le |x| + |0| \le \varepsilon \Leftrightarrow \delta + 0 = \varepsilon \Rightarrow \delta = \varepsilon$. Note that $\forall \varepsilon > 0 \Rightarrow \delta > 0$, then $\lim_{(x,y) \to (0,0)} x = 0$.

Example 2.13. What about $\lim_{(x,y) \to (0,0)} x = 1$? (Ex. 4).

Proof. $\forall \varepsilon > 0 \ \exists \ \delta > 0$ such that $|x| < \delta$ and $|y| < \delta \Rightarrow |x - 1| < \varepsilon \Leftrightarrow |x| + |-1| < \varepsilon \Leftrightarrow \delta + 1 = \varepsilon \Rightarrow \delta = \varepsilon - 1$. But if we take $\varepsilon = 1/2 > 0 \Rightarrow \delta < 0$, then $\lim_{(x,y) \to (0,0)} x \neq 1$.

Example 2.14. Show that $\lim_{(x,y)\to(2,4)} 3x + 5y = 26$ using the method $\varepsilon - \delta$.

Proof. $\forall \varepsilon > 0 \ \exists \ \delta > 0$ such that $|x - 2| < \delta$ and $|y - 4| < \delta \Rightarrow |3x + 5y - 26| < \varepsilon$. But $|3x + 5y - 26| = |(3(x - 2) + 5(y - 4)| < |3(x - 2)| + |5(y - 4)| = |3| \, |x - 2| + |5| \, |y - 4| < \varepsilon \Leftrightarrow 3\delta + 5\delta < \varepsilon \Leftrightarrow 8\delta = \varepsilon \Rightarrow \delta = \frac{\varepsilon}{8}$, then $\lim_{(x,y)\to(2,4)} 3x + 5y = 26$. Note that $\forall \varepsilon > 0 \Rightarrow \delta > 0$, then $\lim_{(x,y)\to(2,4)} 3x + 5y = 26$.

Theorem 2.1. For the real-valued functions $f, g : U \subset \mathbb{R}^n \to W \subset \mathbb{R}$ and the scalars α, β. If $\exists \lim_{x\to x_0} f(x) = L$ and $\exists \lim_{x\to x_0} g(x) = M$, the following five properties hold: $\lim_{x\to x_0} \alpha f(x) = \alpha L$ $\lim_{x\to x_0} [f(x) + g(x)] = L + M$ $\lim_{x\to x_0} [f(x)g(x)] = LM$ $\lim_{x\to x_0} [\frac{f(x)}{g(x)}] = \frac{L}{M}$, if $M \neq 0$ $\lim_{x\to x_0} [f(x)]^{\frac{r}{s}} = L^{\frac{r}{s}}$, if r and s are integers with no common factors and $s \neq 0$

Example 2.15. Let $f : \mathbb{R}^2 \to \mathbb{R}, x^4 - y^4 + 2x - 3y - 34$. Compute $\lim_{(x,y)\to(1,1)} f(x)$.

Solution 2.15. Function f is the sum of the five functions and the limit of the functions exist (Thm. 4). Therefore, $\lim_{(x,y)\to(1,1)} x^4 - y^4 + 2x - 3y - 34 = \lim_{(x,y)\to(1,1)} x^4 - \lim_{(x,y)\to(1,1)} y^4 + \lim_{(x,y)\to(1,1)} 2x - \lim_{(x,y)\to(1,1)} 3y - \lim_{(x,y)\to(1,1)} 34 = 1 - 1 + 2 - 3 - 34 = -35$.

Proposition 2.1. The limit of a function f is unique if it exists.

Proof. Suppose $\lim_{(x)\to x_0} f(x) = L_1$ and $\lim_{x\to x_0} f(x) = L_2$, then for every $\varepsilon > 0$ there exist $\delta_1, \delta_2 > 0$, such that $0 < |x - x_0| < \delta_1 \Rightarrow |f(x) - L_1| < \varepsilon/2$ and $0 < |x - x_0| < \delta_2 \Rightarrow |f(x) - L_2| < \varepsilon/2$. Let $\delta = \min(\delta_1, \delta_2) > 0$ then, since $x_0 \in \mathbb{R}$ there exists $0 < |x - x_0| < \delta$. Therefore, $|L_1 - L_2| \leq |L_1 - f(x)| + |f(x) - L_2| < \varepsilon$. Since this holds for arbitrary $\varepsilon > 0$ we must have $L_1 = L_2$.

2.4.1. L' Hospital's Rule

The limit of a real-valued function $f : \mathbb{R}^n \to \mathbb{R}$ converges to the limit of a real-valued function $f : \mathbb{R} \to \mathbb{R}$. For this reason, it is important to revise this rule [26, 27].

Theorem 2.2. Type $\frac{0}{0}$: For I an interval $a \in I$ and $f, g: I\backslash\{a\} \subset \mathbb{R} \to \mathbb{R}$, the functions have to meet these conditions: (i) the derivatives of f and g exist in $I\backslash\{a\}$. (ii) $g'(x) \neq 0 \; \forall x \in I\backslash\{a\}$. (iii) $\lim_{x \to a} f(x) = \lim_{x \to a} g(x) = 0$, then

$$\lim_{x \to a} \frac{f'(x)}{g'(x)} = L \in \mathbb{R} \Rightarrow \lim_{x \to a} \frac{f(x)}{g(x)} = L.$$

Example 2.16. Let the real-valued functions $f(x) = e^x - 1 - x$, $g(x) = x^2$, and a point $a = 0$. Compute the $\lim_{x \to 0} \frac{f(x)}{g(x)}$.

Solution 2.16. The functions f and g comply with (Thm. 4.1, i–iii). So $\lim_{x \to a} \frac{f'(x)}{g'(x)} = \lim_{x \to 0} \frac{e^x - 1}{x^2}_{[\frac{0}{0}]} = \lim_{x \to 0} \frac{e^x}{2x}_{[\frac{0}{0}]} = \lim_{x \to 0} \frac{e^x}{2}_{[\frac{0}{0}]} = \frac{1}{2}$.

Theorem 2.3. Type $\frac{\infty}{\infty}$: For I an interval $a \in I$ and $f, g: I\backslash\{a\} \subset \mathbb{R} \to \mathbb{R}$, The functions have to meet: (i) the derivatives of f and g exist in $I\backslash\{a\}$. (ii) $g'(x) \neq 0 \; \forall x \in I\backslash\{a\}$. (iii) $\lim_{x \to a} f(x) = \pm\infty$, and $\lim_{x \to a} g(x) = \pm\infty$ then

$$\lim_{x \to a} \frac{f'(x)}{g'(x)} = \lim_{x \to a} \frac{f(x)}{g(x)}.$$

Example 2.17. Let the real-valued functions $f(x) = 2x^3$, $g(x) = x^2 + 1$, and a point $a = \infty$. Compute the $\lim_{x \to \infty} \frac{f(x)}{g(x)}$.

Solution 2.17. The functions f and g comply with (Thm. 4.1, i–iii). So $\lim_{x \to a} \frac{f'(x)}{g'(x)} = \lim_{x \to \infty} \frac{2x^3}{x^2 + 1}_{[\frac{\infty}{\infty}]} = \lim_{x \to \infty} \frac{6x^2}{2x}_{[\frac{\infty}{\infty}]} = \lim_{x \to \infty} \frac{12x}{2}_{[\frac{\infty}{\infty}]} = 6\lim_{x \to \infty} x = \infty$.

2.4.2. Limits of Trajectories

Evaluating the limits of different trajectories intercepting limit point x_0, is a useful process to give a deeper understanding of this operator.

From Prop. 4, $f: \mathbb{R}^n \to \mathbb{R}$, $\lim_{x \to x_0} f(x) = L \Leftrightarrow \lim_{x^+ \to x_0} f(x) = L = \lim_{x^- \to x_0} f(x), L \in \mathbb{R}$; otherwise the limit does not exist. The limit L of the real-valued function f is independent of the trajectory followed to approach point x_0, considering $f: \mathbb{R}^2 \to \mathbb{R}$, whose domain region is (Fig. **2.7**).

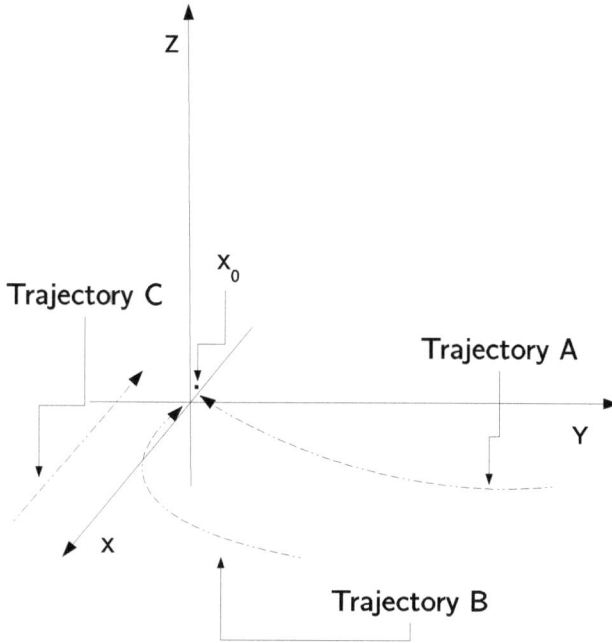

Fig. (2.7). Let the function $f: \mathbb{R}^2 \to \mathbb{R}$ in the xy-plane. The trajectories A and B intercept at point x_0, the trajectory C does **not** reach the point x_0.

Note 2.3. The approximation to point x_0 in the xy-plane accepts an infinite number of trajectories. For this reason, this method is useful to show a limit does not exist.

Example 2.18. Compute the limit $\lim_{x \to 0} \frac{x^2 - y^2}{x^2 + y^2}$ using different trajectories.

Solution 2.18. (i) $(x, 0) \Rightarrow \lim_{(x,0) \to (0,0)} \frac{x^2 - 0^2}{x^2 + 0^2} = \lim_{x \to 0} \frac{x^2}{x^2} = 1$. (ii) $(0, y) \to$ $(0,0) \Rightarrow \lim_{(0,y) \to (0,0)} \frac{0^2 - y^2}{0^2 + y^2} = \lim_{y \to 0} -\frac{y^2}{y^2} = -1$. The limit L in (i) and (ii) is different, therefore, the limit does not exist.

2.5. CONTINUITY

The continuity of a function characterises a function and uses the operator **limit** to determine if a function is defined at any point of its domain.

Definition 2.8. Let $f: U \subset \mathbb{R}^n \to W \subset \mathbb{R}^m$ and $x_0 \in U$. The function f is **continuous** [28] at x_0 if and only if $\lim_{x \to x_0} f(x) = f(x_0)$, otherwise we say that the function is **discontinuous** [25].

Theorem 2.4. Let $f: U \subset \mathbb{R}^n \to W \subset \mathbb{R}^m$ and $x_0 \in U$. Then, f is continuous at $x_0 \in U$ if and only if $\forall \varepsilon > 0$, such that $x_0 \in U$ and $||x - x_0|| < \delta$ imply $||f(x) - f(x_0)|| < \varepsilon$.

Example 2.19. Let the function $f: \mathbb{R}^2 \to \mathbb{R}, x^6 y + 2xy$. If the limit is defined at point $(1,2)$ is function f continuous?

Solution 2.19. $\lim_{(x,y) \to (1,2)} x^6 y + 2xy = 6$ and $f(1,2) = 6$, then function $x^6 y + 2xy$ is continuous at point $(1,2)$.

Theorem 2.5. Let the continuous real-valued functions $f, g: U \subset \mathbb{R}^n \to W \subset \mathbb{R}^m$ and the scalars α, β. If $\exists \lim_{x \to x_0} f(x) = L$ and $\exists \lim_{x \to x_0} g(x) = M$, then the following five properties hold:

Property 1. $\lim_{x \to x_0} \alpha f(x) = \alpha f(x_0)$

Property 2. $\lim_{x \to x_0} [f(x) + g(x)] = f(x_0) + g(x_0)$

Property 3. $\lim_{x \to x_0} [f(x)g(x)] = f(x_0)g(x_0)$, if $M = 1$

Property 4 . $\lim_{x \to x_0} [\frac{1}{f(x)}] = \frac{1}{f(x_0)}$, if $\forall x \in U$ such that $f(x_0) \neq 0$

Property 5. $\lim_{x \to x_0} [f(x)]^{\frac{r}{s}} = f(x_0)^{\frac{r}{s}}$, if r and s are integers with no common factors and $s \neq 0$

Note 2.4. If functions f and g are continuous at point x_0, the composition function $f \circ g$ is continuous at point x_0.

2.6. DIFFERENTIATION

From a geometrical point of view, a function is differentiable if the proximity between the tangent plane and the surface in the **neighbourhood** of a specific point is short enough.

Definition 2.9. Let $f: \mathbb{R}^2 \to \mathbb{R}$ be a **continuous** real-valued function. f is continuously differentiable if, and only if, the partial derivative functions $\frac{\partial f}{\partial x_i}, i = 1, \cdots, n$ **exist** and are **continuous**.

Definition 2.10. The **partial derivative** of a function $f: \mathbb{R}^n \to \mathbb{R}$ (Eq. 2.6) is its derivative with respect to one of the variables, when the others are held constant.

$$\frac{\partial f}{\partial x} = \lim_{h \to 0} \frac{f(x-h) - f(x)}{h} \tag{2.6}$$

Example 2.20. Let the function $f(x, y) = xy^2 - x^2 y$. What is the partial derivative of the function with respect to $x_0 = (1,2)$?

Solution 2.20.

$$
\begin{aligned}
\frac{\partial f}{\partial x} &= \lim_{h \to 0} \frac{f(x+h) - f(x)}{h} \\
&= \lim_{h \to 0} \frac{(x+h)y^2 - y(x+h)^2 - xy^2 - x^2 y}{h} \\
&= \lim_{h \to 0} \frac{xy^2 + hy^2 - yx^2 - 2xhy - yh^2 - xy^2 + x^2 y}{h} \\
&= \lim_{h \to 0} y^2 - 2xy + yh \\
&= y^2 - 2xy.
\end{aligned}
$$

Then $\dfrac{\partial f}{\partial x}_{(x,y)=(1,2)} = 0$

Example 2.21. Is function f [29] differentiable at point $(0,0)$?

$$
f(x,y) = \begin{cases} \dfrac{xy}{\sqrt{x^2+y^2}} & if \ (x,y) \neq (0,0) \\ 0 & otherwise \end{cases}
$$

Solution 2.21. No, it is not differentiable because its $\dfrac{\partial f}{\partial x}(x, y)$ is not continuous at point $(0,0)$.

Proof. $\dfrac{\partial f}{\partial x}(x, y) = \dfrac{y^3}{(x^2+y^2)^{\frac{3}{2}}}, \dfrac{\partial f}{\partial x}(0, y) = sign(y).$

Example 2.22. Is function f [29] differentiable at point $(0,0)$?

$$f(x,y) = \begin{cases} \dfrac{x^2 y}{\sqrt{x^6 + y^2}} & \textit{if } (x,y) \neq (0,0 \\ 0 & \textit{otherwise} \end{cases}$$

Solution 2.22. No, it is not differentiable because f is not continuous.

Proof. $\lim_{(x,y) \to (0,0)} \dfrac{x^2 y}{\sqrt{x^6 + y^2}} = \lim_{(x,x^3) \to (0,0)} \dfrac{x^5}{\sqrt{x^6 + (x^3)^2}} = \dfrac{x^2}{\sqrt{2}} = 0.$

Example 2.23. Let the function $f(x,y,z) = x\sin x^2 y$. (i) What is its partial derivative with respect to variable x? (ii) What is its partial derivative with respect to variable y? (iii) What is its partial derivative with respect to variable z?

Solution 2.23. (i) $\dfrac{\partial f}{\partial x} = \sin x^2 y + 2x^2 y \cos x^2 y$. (ii) $\dfrac{\partial f}{\partial y} = x^3 \cos x^2 y$. (iii) $\dfrac{\partial f}{\partial z} = 0$.

Definition 2.11. A **function exists** at point x_0, if $f(x_0)$ belongs to the image of function f, otherwise the function does **not exist**.

Example 2.24. Does function $f: \mathbb{R}^2 \to \mathbb{R}, \dfrac{1}{x-3}$ exist at $x_0 = (3,0)$?

Solution 2.24. No, it does not. Function f at $(3,0)$ is $\dfrac{1}{0}!$; it does not belong to $\mathbb{R}$ (image of f).

Definition 2.12. The partial derivatives of a function $f: U \subset \mathbb{R}^n \to W \subset \mathbb{R}^m$ (Eq. 2.7), is the $m \times n$ **matrix of the partial derivatives of f** at point x_0

$$D\, f(x_0) = \begin{pmatrix} \dfrac{\partial f_1}{\partial x_1} & \cdots & \dfrac{\partial f_1}{\partial x_n} \\ \vdots & \ddots & \vdots \\ \dfrac{\partial f_m}{\partial x_1} & \cdots & \dfrac{\partial f_m}{\partial x_n} \end{pmatrix}_{x_0} \tag{2.7}$$

Example 2.25. What is the matrix of the partial derivatives of $f: \mathbb{R}^3 \to \mathbb{R}^2, (e^{2x+3y} + y, xy^3 z)$, at point $x_0 = (1,2,3)$?

Solution 2.25. The matrix of partial derivatives $D\, f(x_0)$ of function f is

$$D f(x_0) = \begin{pmatrix} \dfrac{\partial f_1}{\partial x} & \dfrac{\partial f_1}{\partial y} & \dfrac{\partial f_1}{\partial z} \\[2ex] \dfrac{\partial f_2}{\partial x} & \dfrac{\partial f_2}{\partial y} & \dfrac{\partial f_2}{\partial z} \end{pmatrix}_{x_0=(1,2,3)}$$

$$= \begin{pmatrix} \dfrac{\partial e^{2x+3y}+y}{\partial x} & \dfrac{\partial e^{2x+3y}+y}{\partial y} & \dfrac{\partial e^{2x+3y}+y}{\partial z} \\[2ex] \dfrac{\partial xy^3 z}{\partial x} & \dfrac{\partial xy^3 z}{\partial y} & \dfrac{\partial xy^3 z}{\partial z} \end{pmatrix}_{(1,2,3)}$$

where

$$\dfrac{\partial e^{2x+3y}+y}{\partial x}\bigg|_{(1,2,3)} = (e^{2x+3y}+y)'_x = [2e^{2x+3y}+0]_{(1,2,3)} = 2e^8;$$

$$\dfrac{\partial e^{2x+3y}+y}{\partial y}\bigg|_{(1,2,3)} = (e^{2x+3y}+y)'_y = [3e^{2x+3y}+1]_{(1,2,3)} = 3e^8+1;$$

$$\dfrac{\partial e^{2x+3y}+y}{\partial z}\bigg|_{(1,2,3)} = (e^{2x+3y}+y)'_z = [0e^{2x+3y}+0]_{(1,2,3)} = 0;$$

$$\dfrac{\partial xy^3 z}{\partial x}\bigg|_{(1,2,3)} = (xy^3 z)'_x = [y^3 z]_{(1,2,3)} = 2^3 3;$$

$$\dfrac{\partial xy^3 z}{\partial y}\bigg|_{(1,2,3)} = (xy^3 z)'_y = [3xy^2 z]_{(1,2,3)} = 2^2 9;$$

$$\dfrac{\partial xy^3 z}{\partial z}\bigg|_{(1,2,3)} = (xy^3 z)'_z = [xy^3]_{(1,2,3)} = 2^3.$$

$$D f(x_0) = \begin{pmatrix} 2e^8 & 3e^8+1 & 0 \\[1ex] 24 & 36 & 8 \end{pmatrix}_{x_0=(1,2,3)}$$

Theorem 2.6. For the real-valued function $f : U \subset \mathbb{R}^n \to W \subset \mathbb{R}^m$ and $g : U \subset \mathbb{R}^n \to W \subset \mathbb{R}^p$, whose matrix of partial derivatives $D f(x_0)$ and $D g(x_0)$ are C^1 at point x_0, then the following four properties hold:

Property 1. $D\,(\alpha f)(x_0) = \alpha D f(x_0)$, if $\alpha \in \mathbb{R}$

Property 2. $D (f + g)(x_0) = D f(x_0) + D g(x_0)$

Property 3. $D (f g)(x_0) = D f(x_0)D g(x_0)$

Property 4. $D (f \circ g)(x_0) = D f(y_0)D g(x_0)$, where $y_0 = g(x_0)$

Example 2.26. Let the function $f(x) = 3x^3$. Compute $D (f)(x_0)$ at point $x_0 = 2$.

Solution 2.26.

$$D f(x_0) = \left(\frac{df}{dx}\right)_{x_0=2} = (9x^2)_{x_0=2} = 36$$

2.6.1. Gradient

For the real-valued function $f: U \subset \mathbb{R}^n \to W \subset \mathbb{R}$, whose matrix of partial derivatives $D (f)(x)$ exists, and a point $x_0 \in U$, the operator gradient ∇f (Eq. 2.8) is the vector

$$grad\, f(x_0) = \nabla f = (\frac{\partial f}{\partial x_1}, \frac{\partial f}{\partial x_2}, \cdots, \frac{\partial f}{\partial x_n}) \subset U \qquad (2.8)$$

Example 2.27. Let the function $f: \mathbb{R}^3 \to \mathbb{R}, 3xy - z$. Compute the $grad f$ at point $x_0 = (-1,1,1)$

Solution 2.27. $grad\, f(-1,1,1) = (\frac{\partial f}{\partial x}, \frac{\partial f}{\partial y}, \frac{\partial f}{\partial z})_{x_0=(-1,1,1)} = (3y, 3x, -1)_{x_0=(-1,1,1)} = (3, -3, -1)$.

Definition 2.13. For the function $f: U \subset \mathbb{R}^2 \to W \subset \mathbb{R}$, whose matrix of partial derivatives $D (f)(x)$ exists, and a point $x_0 \in U$, the linear approximation of function f (Eq. 2.9) is defined by

$$L(x) = f(x_0) + \frac{\partial f}{\partial x_{x_0}} (x - x_0) + \frac{\partial f}{\partial y_{x_0}} (y - y_0) \qquad (2.9)$$

Note 2.5. The **linear approximation** of a function f plays an important role in finding higher-order polynomial approximations of functions (Sect. 2.7).

Example 2.28. Let the function $f: U \subset \mathbb{R}^2 \to W \subset \mathbb{R}, xy$. Compute the linear approximation at point $x_0 = (1,3)$.

Solution 2.28.

$$L(x)_{x_0=(1,3)} = f(1,3) + \frac{\partial f}{\partial x}_{(1,3)} (x-1) + \frac{\partial f}{\partial y}_{(1,3)} (y-3) = 3 + \frac{\partial(xy)}{\partial x}_{(1,3)} (x-$$

$$1) + \frac{\partial(xy)}{\partial y}_{(1,3)} (y-3) = 3 + 3(x-1) + (y-3) = 3x + y - 3, \qquad \text{then}$$

$$L(x)_{x_0=(1,3)} = 3x + y - 3$$

Theorem 2.7. For the function $f: U \subset \mathbb{R}^n \to W \subset \mathbb{R}$, the vector $grad f$ is normal to the level surface at point x_0.

Proof. Let $f(x) = k$. The tangent line to the level surface at point x_0 is (Def. 2.2.2) $\nabla f(x_0) \cdot (x - x_0) = 0$, so $grad f$ is perpendicular to the level surface at point x. (Fig. **2.9**).

Tangent line to level surface at point x_0

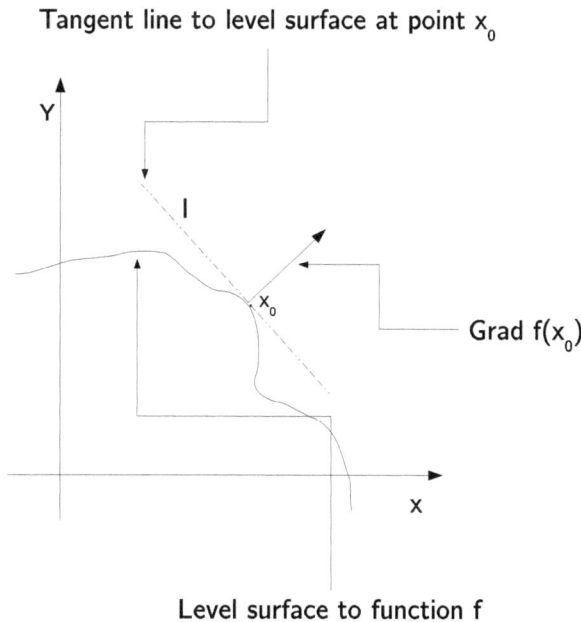

Level surface to function f

Fig. (2.8). The vector $grad f$ is normal to the tangent line l of the level curve at point x_0.

Example 2.29. Let the function $f: \mathbb{R}^2 \to \mathbb{R}, x^2 + y^2$. (i) Compute the level surface $L(c)$ with $c = 4$. (ii) Compute the $grad f$ at point $x_0 = (2,0)$. (iii) Compute the tangent line l to the level surface at point x_0. (iv) Compute the line n at point x_0 with orientation $grad\ f(x_0)$. (v) Demonstrate that lines l and n are perpendicular.

Solution 2.29. (i) $L(c) = L(4) = x^2 + y^2 = 4$ is a circumference with radius $r = 2 \Rightarrow$ at point $x_0 = (2,0)$. (ii) $grad f(2,0) = (\frac{\partial f}{\partial x}, \frac{\partial f}{\partial y})_{x_0=(2,0)}(2x, 2y)_{x_0=(2,0)} = (4,0)$. (iii)

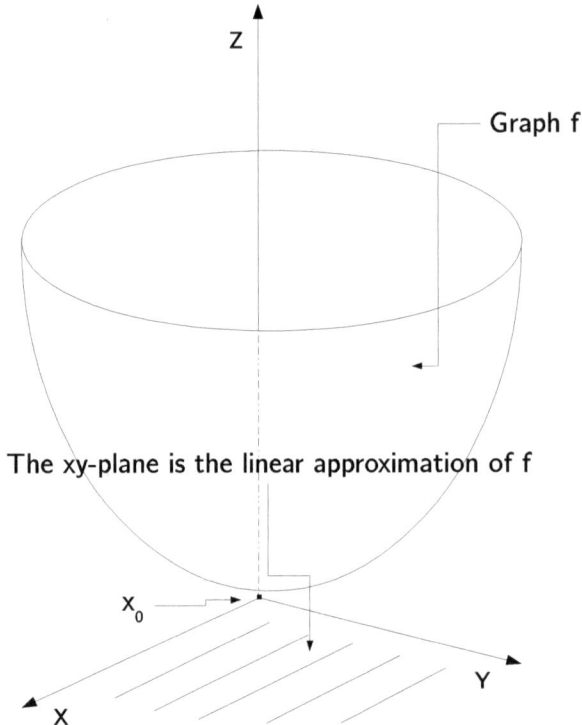

Fig. (2.9). The graph of the linear approximation to function $f: \mathbb{R}^2 \to \mathbb{R}, x^2 + y^2$ in the point $(0,0)$ is the xy-plane.

The tangent line to the level surface at point $x_0 = (2,0)$ is $l(t) = (2,0) + t(0,1), t \in \mathbb{R}$. (iv) The line $n(t) = (2,0) + tgrad\ f(2,0) = (2,0) + t(2x, 2y)_{x_0=(2,0)}, t \in \mathbb{R} = (2,0) + t(4,0), t \in \mathbb{R}$. (v) The orientation vectors of lines $l(t)$ and n are $(0,1)$ and $(4,0)$ respectively. Since the dot product is $(0,1) \cdot (4,0) = 0 \Rightarrow$ the lines l and n are perpendicular.

Theorem 2.8. For the function $f: U \subset \mathbb{R}^n \to W \subset \mathbb{R}$, the $grad f \subset U$ at point x_0 indicates the direction f increases fastest, and the direction of vector $-grad f$ is the direction f decreases fastest.

Proof. The rate of change of f in the direction of any vector v is $||\nabla f(x_0) \cdot v||$, it implies $||\nabla f(x_0)||\cos\theta$, where θ is the angle between v and $\nabla f(x_0)$. The maximum value is when $\theta = 0$, then v and $grad f$ are parallel.

Definition 2.14. For the function $f: U \subset \mathbb{R}^n \to W \subset \mathbb{R}$, the **directional derivative** of f over the line $l(t) = p + tv, t \in \mathbb{R}$ (Eq. 2.10) is

$$Df_v = \frac{d}{dt}\, f \circ l(t)_{t=0} = \frac{d}{dt}\, f(l(t))_{t=0} grad f(l(t)) \cdot l'(t))_{t=0} \subset \mathbb{R} \qquad \textbf{(2.10)}$$

Example 2.30. Let $f: \mathbb{R}^2 \to \mathbb{R}, x^2 + y^3$. Compute the directional derivative of f in the direction of the unit vector $v = (1,1)$ at point $x_0 = (3,4)$.

Solution 2.30. The unit vector v is the vector $\frac{v}{||v||} = \frac{1}{\sqrt{2}}(1,1)$. The **directional derivative** is $Df_v = grad f \cdot v = (2x, 3y^2)_{(3,4)} \cdot \frac{1}{\sqrt{2}}(1,1) = (6,48) \cdot (\frac{1}{\sqrt{2}}, \frac{1}{\sqrt{2}}) = \frac{54}{\sqrt{2}}$

Note 2.6. The existence of all directional derivatives implies the existence of partial derivatives. But the converse is not true.

2.7. POLYNOMIAL APPROXIMATION

Theorem 2.9. For $f: \mathbb{R}^n \to \mathbb{R}$, whose partial derivatives exist and are continuous, the Higher-order Taylor polynomial approximation g (Eq. 2.11) at point x_0 is

$$
\begin{aligned}
g(x) = \quad & f(x_0) + \\
& \sum_{i=1}^{n} \frac{\partial f}{\partial x_i}(x_i - x_{i_0}) + \\
& \frac{1}{2!}\sum_{i=1}^{n}\sum_{j=1}^{n}\frac{\partial^2 f}{\partial x_i \partial x_j}(x_i - x_{i_0})(x_j - x_{j_0}) + \\
& \frac{1}{3!}\sum_{i=1}^{n}\sum_{j=1}^{n}\sum_{k=1}^{n}\frac{\partial^3 f}{\partial x_i \partial x_j \partial x_k}(x_i - x_{i_0})(x_j - x_{j_0})(x_k - x_{k_0})
\end{aligned}
\qquad \textbf{(2.11)}
$$

Example 2.31. Let the function $f: \mathbb{R}^2 \to \mathbb{R}, x^2 + y^2$. Compute the linear approximation at point $x_0 = (0,0)$ using Taylor's approximation.

Solution 2.31.

$$g(x) = f(x_0) + \sum_{i=1}^{n}\frac{\partial f}{\partial x_i}(x_i - x_{i_0})$$

$$= f(0,0) + \frac{\partial f}{\partial x}(x - 0) + \frac{\partial f}{\partial y}(y - 0)$$

$$= 0 + 0(x - 0) + 0(y - 0)$$

The linear approximation of f is $g(x, y) = 0$, *i.e.* it is the xy-plane (Fig. **2.9**).

2.8. CASE STUDY: EQUISPACED LEVEL CURVES

Case 2.1. Identify on the equispaced level curves (Fig. **8**) of a function $f(x, y)$. (i) Where ∇f is largest?; (ii) Where ∇f is smallest?; (iii) Where is $\partial_x f = 0$?; (iv) Where the directional derivative $D_{(i/\sqrt{3}+j/\sqrt{3})} f = 0$. Recall that the level curves of f are curves where f has a constant value. See (Fig. **2.10**). **Case adapted with permission of the author** [3]).

(i) The only information we have is a sketch of the level curves of f, therefore, we need to relate these level curves to the magnitude of ∇f.

Note 2.7. The magnitude of ∇f at a point (x, y) is the directional derivative in the direction of steepest ascent.

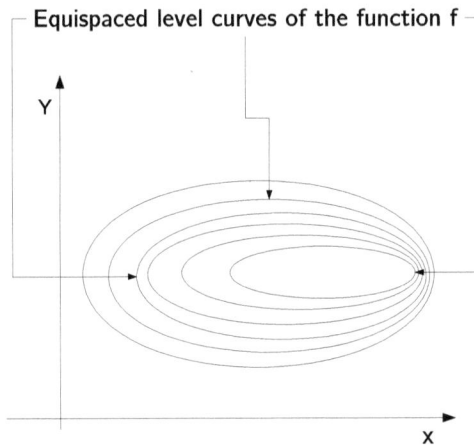

Fig. (2.10). Equispaced level curves of the function f, whose graph is in $\mathbb{R}^3$.

We now relate the level curves to the directional derivatives of f. The level curves correspond to the equally spaced values of f, therefore, the directional derivative is largest where the level curves are nearest. Thus, the maximal magnitude $|\nabla f|$ is located as shown (Fig. **2.11**).

(ii) The magnitude of ∇f is small at points where f is not very steep in any direction. The smallest possible value $|\nabla f|$ can take is 0. It occurs if the function is locally flat. Function f must have a gradient 0 somewhere in that range. Hence $|\nabla f|$ is smallest at such a point (Fig. **2.12**).

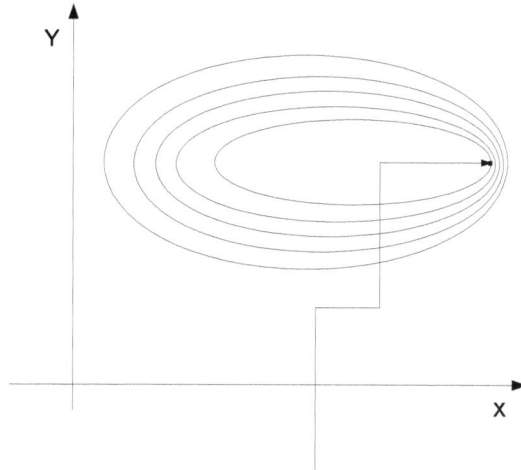

Maximal magnitude is the max |grad f|

Fig. (2.11). The **maximal magnitude** $|\nabla f|$ is located where the level curves are nearest.

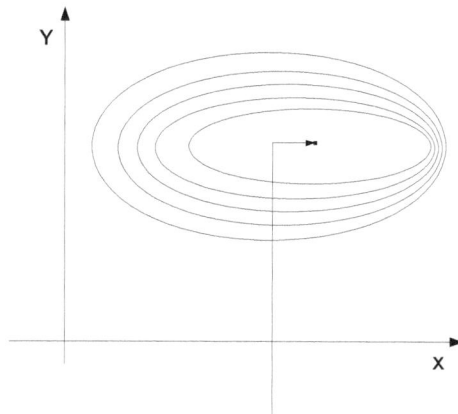

Minimal magnitude is the min |gradient f|

Fig. (2.12). The **minimal magnitude** $|\nabla f|$ is located where $|\nabla f| = 0$.

(iii) We need to relate the partial derivative of f to the level sets of f. Some points where the level curves are parallel to i are as shown (Fig. **2.13**).

Note 2.8. $\partial_x f = 0$ means that the function f is not changing in the x direction. Hence, the level curves of f are parallel to i at these points.

Some points where the level curves are
parallel to vector i where $\partial_x f = 0$

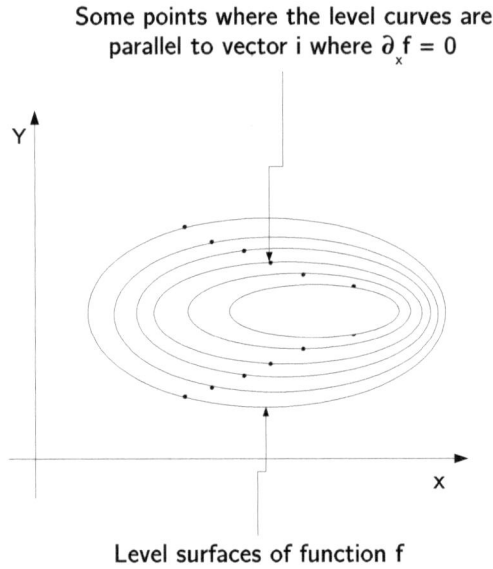

Level surfaces of function f

Fig. (2.13). Points where the level curves are parallel to i, *i.e.* $\partial_x f = 0$.

(iv) We need to relate the directional derivative of f to the level sets of f. Then $D_{(i/\sqrt{3}+j/\sqrt{3})}$

$f = 0$ means that function f is not changing in the direction of $\mathbf{i}/\sqrt{3} + \mathbf{j}/\sqrt{3}$. That is, the level curve is parallel to $i + j$. Some points where the level curves are parallel to $i + j$ are as shown (Fig. **2.14**).

Points where the level curves
are parallel to the vector i + j

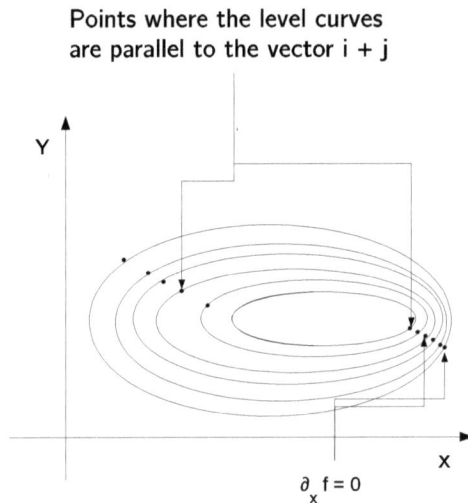

$\partial_x f = 0$

Fig. (2.14). Points where the level curves are parallel to $i + j$.

2.9. CASE STUDY: MARGINAL DEMAND

Case 2.2. Suppose there are two products in the market that are closely related. The change of price in the first product affects its own demand and the demand of the second product. Bearing this in mind, we can think that the function of the demand of product 1 depends on its own price as well as the price of product 2 and *vice versa*. **Case adapted with permission of the author** [4]. So

$$q_1 = f(p_1, p_2)$$

$$q_2 = f(p_1, p_2)$$

If these functions are derivable with respect to p_1 y p_2, these derivatives are called marginal demand

$$\frac{\partial q_i}{\partial p_j} \text{ is the marginal demand of } q_i \text{ with respect to } p_j.$$

So, if the price of product 1 p_1 increases then the demand for product q_1 decreases. If there is a suitable marginal demand, we have

$$\frac{\partial q_i}{\partial p_j} < 0.$$

There can be different scenarios in which these two products can be related, one of them being **two complementary products** where the increase in price of one causes the decrease in demand of the other.

Stating the above

Definition 2.15. Two products are complementary [4] if

$$\frac{\partial q_1}{\partial p_2} < 0 \text{ and } \frac{\partial q_2}{\partial p_1} < 0$$

An example of two complementary products would be oil and gasoline. Most likely, if the price of oil goes up the price of gasoline will rise and the demand for gasoline will decrease.

Another type of relationship between two products is when the increase in price of one product leads to the increase in demand for another. In this case, we refer to a product that can compete with another or substitute it. An example would be fish and meat; meat substitutes fish as a source of protein when the price of fish goes up. Similarly, if the price of meat rises people tend to buy more fish.

Marginal demand can define products that compete with others or substitute them.

Definition 2.16. We say that two products compete with or substitute each other if [4]

$$\frac{\partial q_1}{\partial p_2} > 0 \text{ and } \frac{\partial q_2}{\partial p_1} > 0$$

The equations for the demand of two related products are given by

$$q_1 = 100 - 0.02p_1 - 0.001p_2$$

$$q_2 = 750 + \frac{2}{p_1 + 4} + \frac{3}{p_2 + 2}$$

Using partial derivatives, determine if the products compete with each other, are complementary or neither of them.

We calculate the partial derivatives

$$\frac{\partial q_1}{\partial p_2} \text{ and } \frac{\partial q_2}{\partial p_1}$$

$$\frac{\partial q_1}{\partial p_2} = -0.001$$

$$\frac{\partial q_2}{\partial p_1} = -\frac{2}{(p_1 + 4)^2}$$

Since $(p_1 + 4)^2$ for every possible value of p_1, then $\frac{\partial q_2}{\partial p_1} < 0$ for any value of p_1. Since $\frac{\partial q_1}{\partial p_2} < 0$, then the products are complementary.

2.10. EXERCISES

Exercise 2.1. Let function $f(x, y) = \frac{xy}{x+y}$. (i) What is the limit of function f at point $(0,0)$? (ii) What is the limit of function f at point $(5,1)$ [30].

Exercise 2.2. Let function $f(x, y) = \frac{x^3 y}{x^6 + y^2}$. (i) What is the limit of function f at point $(0,0)$?

Exercise 2.3. Find the directional derivative of function $f(x, y, z) = xyz^3$ in the direction of vector $v = (3, -3, 2)$ [31].

Exercise 2.4. Find the matrix of partial derivatives of function $r(t) = (2t + 1, 3t, 1 - t)$ at point $t_0 = 1$.

Exercise 2.5. Find function $f \circ g$, where $f(t) = (t^3, 3t, 3t + 1)$ and $g(x, y) = (2x - y, xy)$

Exercise 2.6. Suppose the depth of the ocean at point (x, y) is given by $h(x, y) = xy$ and the distance to get there is given by function $p(t) = (t, \frac{t}{2})$. Then your depth time t is given by function $h \circ p(t)$. (i) What is $h \circ p$? (ii) What is the rate of change of depth at time t?

Exercise 2.7. Find the derivative matrix of function $f(x, y, z) = (xy^2, e^x \sin xy, x \ln yx)$ at point $x_0 = (2, 1, 2)$.

CHAPTER 3

Critical Points

Abstract: This chapter focuses on the identification of the maximum, minimum, or saddle points located in the domain of the real-valued function in a graph of a function. The first part of this chapter focuses on the open set domain of a function and the second part on the closed set domain. The characterization of these points called critical points" is resolved with the Hessian matrix and the Bordered Hessian matrix. Finally, we will review the Implicit Function Theorem.

Keywords: Bordered Hessian matrix, Closed set, Critical points, Hessian matrix, Implicit function theorem, Inverse function theorem, Inflection value, Inverse function, Maximum points, Maximum value, Minimum points, Minimum value, Open set, Saddle point, Second order partial derivative.

3.1. PRELIMINARIES

The related operators would be setting of the **real coordinate space** $\mathbb{R}^n$, but cases and examples will be displayed on the plane and space for a better understanding of the concepts.

3.2. POSSIBILITIES AND RESTRICTIONS

The first part of this chapter describes a set of methods that resolve functions of several variables in a restricted domain, identifying and describing the critical points as maximum or minimum. These techniques are useful to maximise or minimise functions that present application problems.

The second part describes and verifies the Implicit Function Theorem. This theorem is certainly the most important theorem in calculus for the impact it has on implementing problems. It enables the identification of the subset in the domain of the function, where the function is differentiable and thus invertible, and it analytically determines its partial derivatives. Although this theorem does not explicitly define the inverse function, it makes it possible to construct a polynomial approximation (sect. 2.7) from the partial derivatives found.

This theorem makes possible the relation between vector spaces, providing a bijection in a subset of the domain.

3.3. OPEN SET DOMAINS

The methods used to identify whether the points located in the domain of a real-valued function f are **critical points**, as well as the methods used to decide if those critical points are maximum, minimum, or saddle points, will depend on whether the **domain** of the function is open or closed.

The domain of a function f can be closed or open, depending on whether the set that forms its domain is open or closed.

Definition 3.1. A set U is an open set if for all $x \in U$ there exists an open ball with radius $\varepsilon > 0$, such that $B(x, \varepsilon) \subset U$ (Fig. **3.1**).

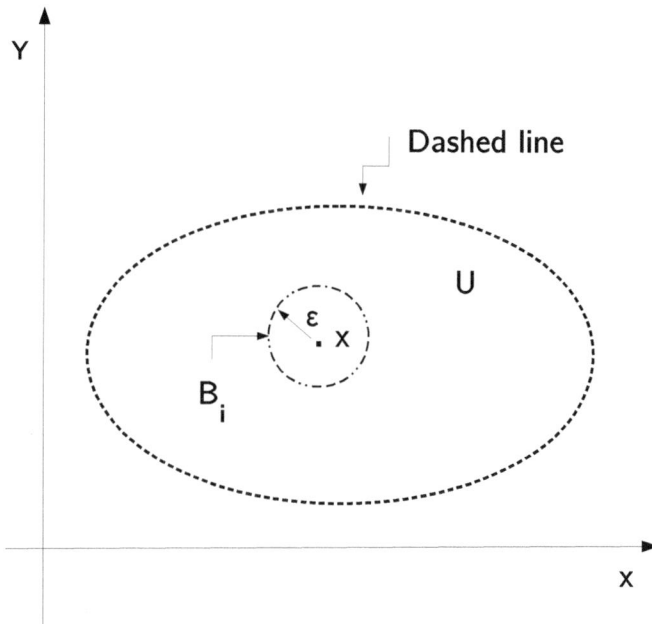

Fig. (3.1). If for all x in U exists open ball $B_i \subset U$, then the set U is open.

Example 3.1. Let the real-valued function $f : \mathbb{R}^2 \to \mathbb{R}, x^3 y^2$. Is its domain open or closed?

Solution 3.1. The space $\mathbb{R}^2$ is an open set, therefore, its domain is open.

3.3.1. Critical Points in Open Sets

Definition 3.2. The **critical points** x_0 of a function f are points located in the open domain where $\nabla f(x_0) = 0$ (Def. 2.12). These points are located where the graph of function f reaches a **maximum**, a **minimum** or **changing concavity**. The latter are points of the **saddle type**), *i.e.* when the graph of function f before point x_0 is concave upward and after this point it is concave downward, or *vice versa* (Fig. **3.2**). These points must be evaluated by the **Hessian matrix** to know if they are maximum, minimum, or saddle points.

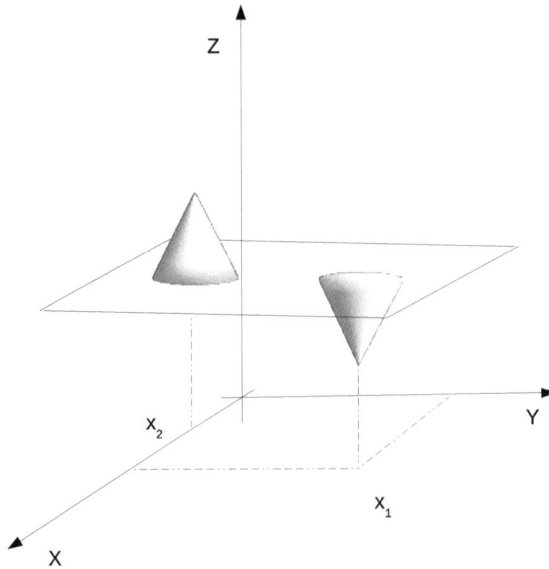

Fig. (3.2). The point x_1 is a minimum critical point, while the point x_2 is a maximum critical point.

Example 3.2. Let function $f: \mathbb{R}^2 \to \mathbb{R}, x^2 + y^2$ (Fig. **2.1**). Compute the critical points (if any).

Solution 3.2. By $\nabla f(x) = (0,0)$ we obtain only one critical point $(0,0)$.

$$\nabla f(x) = \left(\frac{\partial x^2 + y^2}{\partial x}, \frac{\partial x^2 + y^2}{\partial y} \right) = (2x, 2y) = (0,0)$$

Example 3.3. Let function $f: \mathbb{R}^3 \to \mathbb{R}, 1$. Compute the critical points (if any).

Solution 3.3. By $\nabla f(x) = (0,0,0)$ all points $x \in \mathbb{R}^3$ are critical points.

$$\nabla f(x) = \left(\frac{\partial 1}{\partial x}, \ \frac{\partial 1}{\partial y}, \ \frac{\partial 1}{\partial z} \right) = (0,0,0) = (0,0,0)$$

Example 3.4. Let function $f: \mathbb{R}^2 \rightarrow \mathbb{R}, x^3 + 3x^2 - 9x + y^3 - 12y$ [32]. Compute the critical points (if any).

Solution 3.4. By $\nabla f(x) = (0,0)$, it is implied that $3x^2 + 6x - 9 = 0 \ 3y^2 - 12 = 0$, therefore, the critical points are $(1,2), (1,-2), (-3,2),$ and $(-3,-2)$.

$$\nabla f(x) = \left(\frac{\partial x^3 + 3x^2 - 9x + y^3 - 12y}{\partial x}, \ \frac{\partial x^3 + 3x^2 - 9x + y^3 - 12y}{\partial y} \right) = (0,0)$$

3.3.2. Hessian Matrix

Definition 3.3. For the function $f: (x_1, x_2, \cdots, x_n) \in \mathbb{R}^n \rightarrow W \subset \mathbb{R}$, in order to know if a critical point x_0 is maximum, minimum or a saddle point we have to: (i) evaluate the Hessian matrix $H(f(x))$ (Eq. 3.1) at the critical point x_0; (ii) construct a succession of determinants $H_i \ i = 1, \cdots, n$ whose elements are taken from the Hessian matrix H; (iii) evaluate every determinant H_i at the critical point. If the pattern of succession is $+, +, +, \cdots, +$ the critical point x_0 is **minimum**. If the pattern of succession is $-, +, -, \cdots, +$ the critical point x_0 is **maximun**. If $H_i \neq 0$ the critical point x_0 is where the **concavity changes** and it is a **saddle point**.

$$H(f(x)) = \begin{pmatrix} \frac{\partial^2 f}{\partial x_1 \partial x_1} & \frac{\partial^2 f}{\partial x_1 \partial x_2} & \cdots & \frac{\partial^2 f}{\partial x_1 \partial x_n} \\ \frac{\partial^2 f}{\partial x_2 \partial x_1} & \frac{\partial^2 f}{\partial x_2 \partial x_2} & \cdots & \frac{\partial^2 f}{\partial x_2 \partial x_n} \\ \vdots & \vdots & \ddots & \vdots \\ \frac{\partial^2 f}{\partial x_n \partial x_1} & \frac{\partial^2 f}{\partial x_n \partial x_2} & \cdots & \frac{\partial^2 f}{\partial x_n \partial x_n} \end{pmatrix}_{x_0} \quad \textbf{(3.1)}$$

Note 3.1. The **second order partial derivative** $\frac{\partial^2 f}{\partial y \partial x}$ is equivalent to $\frac{\partial f}{\partial y}[\frac{\partial f}{\partial x}]$, *i.e.* first we derive function f with respect to x and then with respect to y.

Note 3.2. If any of the determinants evaluated at the critical point is **zero**, the method does not apply. In this case, it is necessary to evaluate the function in a neighbourhood of the critical point.

Example 3.5. Let the function $f: \mathbb{R}^2 \to \mathbb{R}, x^2 + y^2$ (Fig. **2.1**), whose unique critical point is $(0,0)$. What kind of critical point is it?

Solution 3.5. The pattern of succession is $+,+$, so the critical point $(0,0)$ is minimum.

$$H_1(f(x)) = \left[\frac{\partial^2 x^2 + y^2}{\partial x\, \partial x} \right]_{x_0=(0,0)} = [2]_{x_0=(0,0)} = 2 > 0$$

$$H_2(f(x)) = \begin{vmatrix} \dfrac{\partial^2 x^2 + y^2}{\partial x\, \partial x} & \dfrac{\partial^2 x^2 + y^2}{\partial x\, \partial y} \\ \dfrac{\partial^2 x^2 + y^2}{\partial y\, \partial x} & \dfrac{\partial^2 x^2 + y^2}{\partial y\, \partial y} \end{vmatrix}_{x_0=(0,0)} = \begin{bmatrix} 2 & 0 \\ 0 & 2 \end{bmatrix}_{x_0=(0,0)} = 4 > 0$$

Example 3.6. Let the function $f: x^2 + y^2 + z^2 \le 1 \subset \mathbb{R}^3 \to \mathbb{R}, x^2 + y^2 + z^2$. (i) What is the graph of the function? (ii) What is the graph of the domain of the function? (iii) Compute its critical points. (iv) What kind of critical points are they?

Solution 3.6. (i) The graph of function f does not exist in the Cartesian system because it belongs to the space $\mathbb{R}^4$. (ii) The graph of the domain of this function is a solid unitary sphere. (iii) By $\nabla f(x) = (0,0,0)$ we obtain only one critical point $(0,0,0)$.

$$\nabla f(x) = \left(\frac{\partial x^2 + y^2 + z^2}{\partial x}, \frac{\partial x^2 + y^2 + z^2}{\partial y}, \frac{\partial x^2 + y^2 + z^2}{\partial z} \right) = (2x, 2y, 2z)$$
$$= (0,0,0)$$

(iv) The pattern of succession is $+,+,+$, so the critical point $(0,0,0)$ is minimum.

$$H_1(f(x)) = \left[\frac{\partial^2 x^2 + y^2 + z^2}{\partial x\, \partial x} \right]_{x_0=(0,0)} = [2]_{x_0=(0,0)} = 2 > 0$$

$$H_2(f(x)) = \begin{bmatrix} \dfrac{\partial^2 x^2 + y^2 + z^2}{\partial x\, \partial x} & \dfrac{\partial^2 x^2 + y^2 + z^2}{\partial x\, \partial y} \\[2em] \dfrac{\partial^2 x^2 + y^2 + z^2}{\partial y\, \partial x} & \dfrac{\partial^2 x^2 + y^2 + z^2}{\partial y\, \partial y} \end{bmatrix}_{x_0=(0,0)}$$

$$= \begin{bmatrix} 2 & 0 \\[1em] 0 & 2 \end{bmatrix}_{x_0=(0,0)} = 4 > 0$$

$$H_3(f(x))$$

$$= \begin{bmatrix} \dfrac{\partial^2 x^2 + y^2 + z^2}{\partial x\, \partial x} & \dfrac{\partial^2 x^2 + y^2 + z^2}{\partial x\, \partial y} & \dfrac{\partial^2 x^2 + y^2 + z^2}{\partial x\, \partial z} \\[2em] \dfrac{\partial^2 x^2 + y^2 + z^2}{\partial y\, \partial x} & \dfrac{\partial^2 x^2 + y^2 + z^2}{\partial y\, \partial y} & \dfrac{\partial^2 x^2 + y^2 + z^2}{\partial y\, \partial z} \\[2em] \dfrac{\partial^2 x^2 + y^2 + z^2}{\partial z\, \partial x} & \dfrac{\partial^2 x^2 + y^2 + z^2}{\partial z\, \partial y} & \dfrac{\partial^2 x^2 + y^2 + z^2}{\partial z\, \partial z} \end{bmatrix}_{x_0=(0,0,0)}$$

$$H_3(f(x)) = \begin{bmatrix} 2 & 0 & 0 \\[1em] 0 & 2 & 0 \\[1em] 0 & 0 & 2 \end{bmatrix}_{x_0=(0,0,0)} = 8 > 0$$

3.4. CLOSED SET DOMAINS

The methods used to identify whether the points located in the domain of a real-valued function f are **critical points**, as well as the methods used to decide if those critical points are maximum, minimum, or saddle points, will depend on whether the **domain** of the function is open or closed.

Definition 3.4. A set M is **closed** if it contains all its boundary points [25] in set $M \cup \partial M$ (boundary of M). If the domain of a function f contains any closed set, the domain is closed.

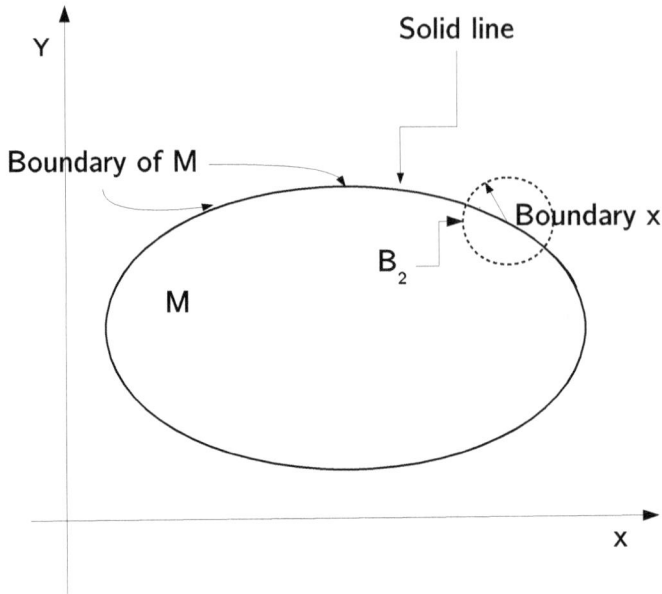

Fig. (3.3). Be the ball $B_2 \not\subset M$. The set M is closed because it includes the set ∂M.

3.4.1. Critical Points in Closed Sets

Definition 3.5. The points x_0 of a function $f: W \subset \in \mathbb{R}^n \to W \subset \mathbb{R}$, where the set W is bounded by the functions $g_{j=1,\cdots m}(x_0) = k, k \in \mathbb{R}$ and comply with $\nabla f(x_0) = \lambda \nabla g(x_0), \lambda \in \mathbb{R}$, could be **critical points**. These points (if any) must be evaluated by the **Bordered Hessian matrix** to know if they are **maximum**, **minimum** or **saddle points**.

Note 3.3. In an open domain, this algorithm looks for points in the open set M, while in a closed domain the algorithm looks for critical points exclusively in the set ∂M. A critical point in the set ∂M does not necessarily correspond with a critical point in the open set. A closed set would be $M \cup \partial M$.

Example 3.7. Let the real-valued function $f(x,y) = 1 - x - y$ represent the top face of a tetrahedron (Fig. **3.4**). (i) If the domain of the function is open, what would be the critical points? (ii) If the domain is closed, *i.e.* a triangle bounded by the vertices $(0, 0, 0)$, $(1, 0, 0)$, and $(0, 1.0)$ (Fig. **3.5**), what would be the critical points?

Solution 3.7. (i) The domain of the function is the open set $\mathbb{R}^2$. In that region there are no points where $\nabla f(x_0) = 0$, therefore, there are no critical points. (ii) Let the closed region $S = \{(x,y) \in \mathbb{R}^2 | y = 1 - x, x \in [0,1]\}$, then $g(x,y) = y + x - 1$.

$\nabla f(x_0) = \lambda \nabla g(x_0) \Rightarrow (-1,-1) = \lambda(1,1)$ and $(x, y) \in S$. The solutions are $\lambda = -1$, where $(x, y) \in S$. $f(x, y) = 1 - x = 0$ (line in xy-plane). The points $(x, 1 - x) \in xy - \text{plane} | x \in [0,1]$ are critical points. Let $f(x, y = 0) = 1 - x$ (line in xz-plane) its maximum value is $x = 0$, then $(0,0)$ is a critical point. Let $f(x = 0, y) = 1 - y$ (line in yz-plane) its maximum value is $y = 0$, then $(0,0)$ is a critical point. So, the critical points are $(\lambda, x, y) = (-1,0,0)$ and $(\lambda, x, y) = (-1, x, 1 - x), x \in [0,1]$.

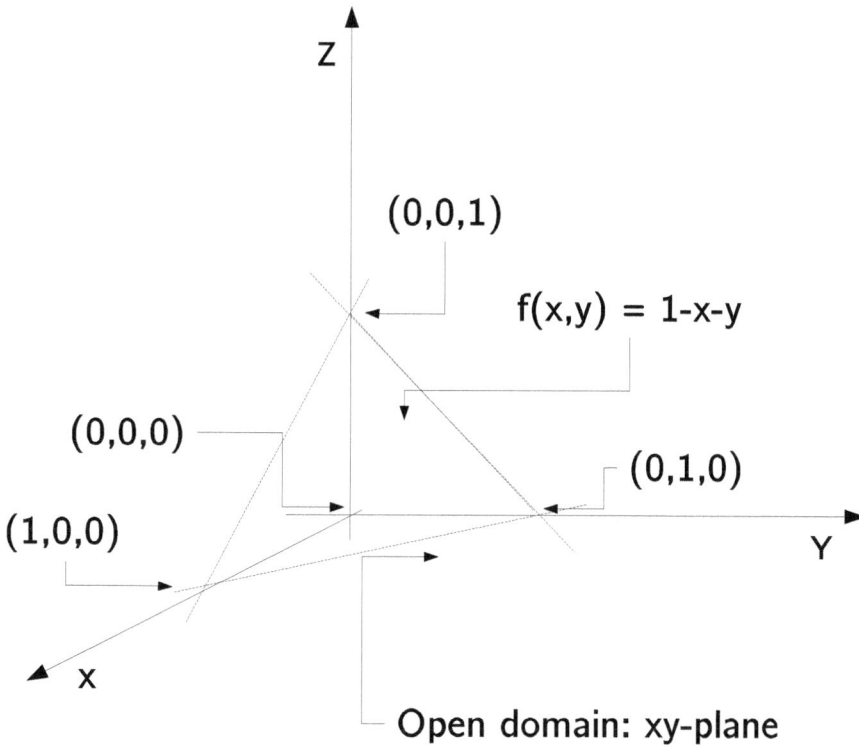

Fig. (3.4). If the domain of the function is open, there are **no** critical points.

Example 3.8. Let the real-valued function $f: S \subset \mathbb{R}^2 \to \mathbb{R}, x^2 + y^2$, where $S = \{(x, y) \in \mathbb{R}^2 | x^2 + y^2 \leq 1, \text{ such that } x, y \in \mathbb{R}\}$. Compute the critical points on the boundary of S.

Solution 3.8. Since $\nabla f(x_0) = \lambda \nabla g(x_0) \Rightarrow (2x, 2y) = \lambda(2x, 2y)$ is restricted to S it implies that (a) $x = 0 \Rightarrow y = \pm 1$ and $\lambda = 1$; (b) $y = 0 \Rightarrow x = \pm 1$ and $\lambda = 1$; (c) $x \neq 0$ and $y \neq 0 \Rightarrow \lambda = 1$ and $x^2 + y^2 = 1$. The critical points are $(\lambda, x, y) = (1, 0, \pm 1)$ and $(1, \pm 1, 0)$.

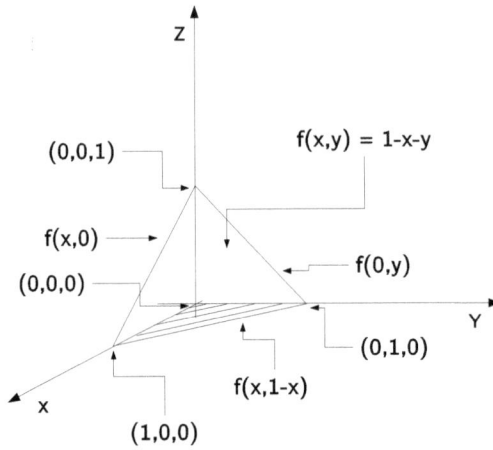

Fig. (3.5). If the domain of the function is closed, there is a **maximum** point at $(0,0,1)$ and a **minimum** point along the line $g(x) = 1 - x$.

3.4.2. Bordered Hessian Matrix

For the function $f: W \subseteq \in \mathbb{R}^n \to W \subset \mathbb{R}$, the closed set W is bounded by the functions $g_{i=1,\cdots m}$, $\lambda \in \mathbb{R}$, where $m < n$ and $h: \mathbb{R}^n \to W \subset \mathbb{R}, f - \lambda g$. To know if a critical point x_0 is maximum, minimum or a saddle point we have to: (i) evaluate the Bordered Hessian matrix $\overline{H}(f(x))$ (Eq. **3.2**) in the critical point x_0; (ii) construct a succession of determinants $\overline{H}_i$ $i = 3, \cdots, n$, whose elements are taken from the Bordered Hessian matrix $\overline{H}$; (iii) evaluate every determinant $\overline{H}_i$ in the critical point. If the elements of the succession are alternated starting at $(-1)^{m+1}$, the critical point is *maximum*. If the elements of the succession are $(-1)^m$, the critical point is *minimum*.

$$\overline{H}(f(\mathbf{x})) = \begin{pmatrix} 0 & \cdots & 0 & \frac{\partial g_1}{\partial x_1} & \cdots & \frac{\partial g_1}{\partial x_n} \\ \vdots & \ddots & \vdots & \vdots & \ddots & \vdots \\ 0 & \cdots & 0 & \frac{\partial g_m}{\partial x_1} & \cdots & \frac{\partial g_m}{\partial x_n} \\ \frac{\partial g_1}{\partial x_1} & \cdots & \frac{\partial g_m}{\partial x_1} & \frac{\partial^2 h}{\partial x_1 \partial x_1} & \cdots & \frac{\partial^2 h}{\partial x_1 \partial x_n} \\ \vdots & \ddots & \vdots & \vdots & \ddots & \vdots \\ \frac{\partial g_1}{\partial x_n} & \cdots & \frac{\partial g_m}{\partial x_n} & \frac{\partial^2 h}{\partial x_n \partial x_1} & \cdots & \frac{\partial^2 h}{\partial x_n \partial x_n} \end{pmatrix}\Bigg|_{x_0}$$

$$(3.2)$$

Note 3.4. If any of the determinants evaluated in the critical point is **zero**, the method does not apply. In this case, it is necessary to evaluate the function in the neighbourhood around the critical point.

Example 3.9. Let the real-valued function $f: S \subset \mathbb{R}^3 \to \mathbb{R}, x^2 + 4y^2$, where $S = \{(x,y) \in \mathbb{R}^2 | x^2 + y^2 = 1, \text{such that } x, y \in \mathbb{R}\}$ [33]. (i) Compute the critical points. (ii) Choose a critical point (if any) and determine whether it is maximum, minimum or a saddle point.

Solution 3.9. (i) The system is $2x = \lambda 2x$, $8y = \lambda 2y$, and $x^2 + y^2 = 1$, then the critical points are $(\lambda, x, y) = (4, 0, \pm 1)$ and $(\lambda, x, y) = (1, \pm 1, 0)$. (ii) Let $f(x,y) = x^2 + 4y^2$ and $g(x,y) = x^2 + y^2 = 1$, since $m = 1$ and $n = 2$ only have one determinant $\overline{H}_3 = 24 > 0$, and $(-1)^{m+1} = (-1)^{1+1} = 1 > 0$ then, the critical point $(\lambda, x, y) = (4, 0, 1)$ is a maximum point.

$$\overline{H}_3(f(x)) = \begin{pmatrix} 0 & \dfrac{\partial g}{\partial x} & \dfrac{\partial g}{\partial y} \\[2mm] \dfrac{\partial g}{\partial x} & \dfrac{\partial^2 h}{\partial x\,\partial x} & \dfrac{\partial^2 h}{\partial x\,\partial y} \\[2mm] \dfrac{\partial g}{\partial y} & \dfrac{\partial^2 h}{\partial y\,\partial x} & \dfrac{\partial^2 h}{\partial y\,\partial y} \end{pmatrix}_{x_0}$$

$$= \begin{pmatrix} 0 & 2x & 2y \\ 2x & 2 - 2\lambda & 0 \\ 2y & 0 & 8 - 2\lambda \end{pmatrix}_{(4,0,1)}$$

$$\overline{H}_3(f(x)) = \begin{bmatrix} 0 & 0 & 2 \\ 0 & -6 & 0 \\ 2 & 0 & 0 \end{bmatrix} = 24 > 0$$

3.5. IMPLICIT FUNCTION THEOREM

The following three cases show the importance of this theorem.

Case A. Let the equation $x + y = 1$. Solving for y we obtain the function $f(x) = 1 - x$. This function is continuous and differentiable.

Case B. Let the equation $x^2 + y^2 = 1$. Solving for y we obtain $f(x) = \pm\sqrt{1 - x^2}$. Note that f **is not** a function but we can build two functions from f: $f_1(x) = \sqrt{1 - x^2}$ representing the upper curve of the circumference and $f_2(x) = -\sqrt{1 - x^2}$ representing the lower curve. The derivative of f **is not** defined at -1 or 1, therefore, the function derived from this equation has problems of differentiability and existence.

Case C. Let the equation $yx^2 + \ln y^2 x + \sin xy = 2$. What is function f? In fact, several questions arise with this equation: does f exist? and if it does, what is f?

An implicit function is defined by an implicit equation. We denote the variable in $\mathbb{R}^{n+1} = \mathbb{R}^n \times \mathbb{R}$ by (x, y), where $x = (x_1, \cdots, x_n)$ is in $\mathbb{R}^n$ and y is in $\mathbb{R}$. **Proofs reproduced with the permission of the author** [34]).

Theorem 3.1. Let an **implicit function** $F: \Omega \to \mathbb{R}$ be of class C^1 in an open set Ω inside $\mathbb{R}^n \times \mathbb{R}$, be (x_0, z_0) a point in Ω such that $F(x_0, z_0) = 0$, and be $\frac{\partial F}{\partial z}(x_0, z_0) > 0$. Thus, there exist the open sets $X \in \mathbb{R}^n$ and $Y \in \mathbb{R}$ with $(x_0, z_0) \in \mathbb{R}^n \times \mathbb{R} \subset \Omega$ that satisfy the following:

(i) For each $x \in X$ there is a unique function $z = f(x) \in Y$ such that $F(x, f(x)) = 0$

(ii) We have $f(x) = z$. Furthermore, $f: X \to Y$ is of class C^1 and

$$\frac{\partial f}{\partial x_j}(x) = -\frac{\frac{\partial F}{\partial x_j}(x, f(x))}{\frac{\partial F}{\partial z}(x, f(x))}, \forall x \in X, \text{where} j = 1, \cdots, n.$$

Proof. [**Existence and Uniqueness.**] Since $\frac{\partial F}{\partial y}(a, b) > 0$, by continuity there exists a non-degenerate $(n + 1)$-dimensional parallelepiped $X' \times [b_1, b_2]$ centred at (a, b) and contained in Ω, whose edges are parallel to the coordinate axes, such that $\frac{\partial F}{\partial y} > 0$ on $X' \times [b_1, b_2]$. Then, the function $F(a, y)$, where y runs over $[b_1, b_2]$, is strictly increasing and $F(a,b) = 0$. Thus, we have $F(a,b_1) < 0$ and $F(a,b_2) > 0$. By the continuity of F, there exists an open non-degenerate n-dimensional parallelepiped X centred at a and contained in X', whose edges are parallel to the coordinate axes such that for every x in X we have $F(x,b_1) < 0$ and $F(x,b_2) > 0$. Hence, fixing an arbitrary x in X and using the intermediate-value theorem on the strictly increasing function $F(x,y)$, where y runs over $[b_1,b_2]$, there exists a unique $y = f(x)$ inside the open interval $Y = (b_1,b_2)$, such that $F(x,f(x)) = 0$.

If $F(x,z) = 0$ and $\frac{\partial F}{\partial z}(x_0,z_0) > 0$, then there exist $F(x_0,z_0 - \varepsilon) < 0$ and $F(x_0 + \varepsilon, z_0) > 0$, $\varepsilon \in \mathbb{R}^+$. Using the intermediate-value theorem on the strictly increasing function $F(x,z)$ there exists a unique $z = f(x)$.

Proof. [**Continuity.**] $\overline{b_1}$ and $\overline{b_2}$ are such that $b_1 < \overline{b_1} < b < b_2 < \overline{b_2}$. Here, there exists an open set X'' contained in X and containing a, such that $f(x)$ is an open interval $(\overline{b_1},\overline{b_2})$ for all x in X''. Thus, f is continuous at $x = a$. Given any a' in X we have $b' = f(a)$, then, $f: X \to Y$ is the solution of the problem $f(x,h(x)) = 0$ for all x in X with the condition $h(a') = b'$. Therefore, f is continuous at a'.

If there is a closed interval $z_0 - \varepsilon < z_0 < z + \varepsilon$ for $x = x_0$ (Thm. 3.2), there exists an open set $X'' \subset X$ such that $f(x) \in X'', \forall x \in X''$. Thus $f(x)$ is continuous at x.

Proof. [**Differentiability.**] Let $F: \Omega \to \mathbb{R}$, $F(x,z) = F(F_1(x), \cdots, F_z(x)) = 0$.

The implicit differentiation of F. $\frac{\partial F}{\partial F_1}\frac{\partial F_1}{\partial x} + \frac{\partial F}{\partial F_2}\frac{\partial F_2}{\partial x} + \cdots + \frac{\partial F}{\partial F_z}\frac{\partial F_z}{\partial x} = \frac{\partial F}{\partial x}\frac{\partial x}{\partial x} +$

$\frac{\partial F}{\partial x_2}\frac{\partial x_2}{\partial x} + \cdots + \frac{\partial F}{\partial z}\frac{\partial z}{\partial x} = \frac{\partial F}{\partial x} 1 + \frac{\partial F}{\partial x_2} 0 + \cdots + \frac{\partial F}{\partial z}\frac{\partial z}{\partial x} = 0$, then $\frac{\partial f}{\partial x} = -\frac{\frac{\partial F}{\partial x}}{\frac{\partial F}{\partial z}}$.

Note 3.5. The theorem states the conditions of existence, uniqueness, and continuity of function f and gives its derivative, but it does not give the function f. Nevertheless, we can use the derivative of f and obtain an equivalent polynomial approximation from Taylor's Theorem (Sec. 2.7).

Example 3.10. Let the function $x^3y^2 - 3xy + 2 = 0$. (i) Is it possible to define function f with respect to the variable x in a neighbourhood of point (1,2)? (ii) What is the derivative of function f?

Solution 3.10. (i) Since $F(1,2) = 0$, F is of class C^1 and $\frac{\partial F}{\partial y} = 2x^3y - 3x \Rightarrow$
$\frac{\partial F}{\partial y}(1,2) = 1 > 0$, then there is a neighbourhood of point $x = 1$. So there exist a unique differential solution $z = f(x)$ with a continuous derivative such that $f(1) = 2$, (ii) $\frac{\partial f}{\partial x} = -\frac{3x^2y^2 - 3y}{2x^3y - 3x}$.

Example 3.11. Lete the function $e^x + e^y + e^z - 3e = 0$. (i) Is it possible to define function f with respect to the variable x, in a neighbourhood of point $(1,1,1)$? (ii) What is the total derivative of function f? (iii) What is a polynomial approximation of function f?

Solution 3.11. (i) Since $F(1,1,1) = 0$, F is of class C^1 and $\frac{\partial F}{\partial z}(1,1,1) = e > 0$, then there is a neighbourhood of point $x = (1,1)$. So there exists a unique differential solution $z = f(x,y)$ and with a continuous derivative such that $f(1,1) = 1$.
(ii) $\frac{\partial f}{\partial x}_{(x_0,y_0)} = -\frac{e^x}{e^z}_{(1,1)} = -1$, and $\frac{\partial f}{\partial y}_{(x_0,y_0)} = -\frac{e^y}{e^z}_{(x_0,y_0)}$

$= -1$. (iii) $f(x,y) = f(x_0,y_0) + \frac{\partial f}{\partial x}_{(x_0,y_0)}(x - x_0) + \frac{\partial f}{\partial y}_{(x_0,y_0)}(y - y_0) + \frac{1}{2}\left[\frac{\partial^2 f}{\partial x\, \partial x}_{(x_0,y_0)}\right.$

$(x - x_0)^2 + \frac{\partial^2 f}{\partial x\, \partial y}_{(x_0,y_0)}(x - x_0)(y - y_0) + \frac{\partial^2 f}{\partial y\, \partial x}_{(x_0,y_0)}(x - x_0)(y - y_0) + \frac{\partial^2 f}{\partial y\, \partial y}_{(x_0,y_0)}$

$(y - y_0)^2\Big] = 1 + x + y - x^2 - y^2$. Particularly, $f(1,1) = 1$.

3.5.1. Inverse Function Theorem

Theorem 3.2. Let a **function** $f: \Omega \to \mathbb{R}$ be of class C^1 in an open set Ω inside $\mathbb{R}^n$ and x_0 a point in Ω, such that $Jf(x)$ is invertible. Then, there exist an open set X containing x_0, an open set Y containing $f(x_0)$, and a function $f^{-1}: Y \to X$. Moreover, the class C^1 satisfies $f(f^{-1}(y)) = y$ for all $y \in Y$ and $f^{-1}(f(x)) = x$ for all $x \in X$. **(Procedure reproduced with the permission of the author [34]).**

$$Jf^{-1}(y) = Jf(f^{-1}(y))^{-1}, \text{for all } y \in Y.$$

Note 3.6. The **Jacobian** determinant $Jf(x)$ of function $f:\mathbb{R}^n \to \mathbb{R}^n$ is the determinant of its matrix of partial derivatives $Df(x)$ (Def. 2.12).

Proof. **[Existence.]** From (Thm. 3.1), there exist $f(f^{-1}(y)) = y, \forall y \in Y$.

Example 3.12. Let the function $f(x) = e^x$. (i) Is it possible to define function f^{-1} with respect to the variable y, in a neighbourhood of point $x = 1$? (ii) Is it differentiable? (iii) What is the function (or polynomial approximation) f^{-1}?

Solution 3.12. (i) Let $f(x) = e^x \to J_{x_0=1}f(e) = |e| = e \neq 0$, then there exists a function f^{-1} of class C^{-1} for function f. (ii) Since $\frac{\partial f}{\partial x}_{x_0=1} = e$ and $\frac{\partial f^{-1}}{\partial y}_{y_0} = \frac{1}{\frac{\partial f}{\partial x}_{x_0}} = \frac{1}{e}$. (iii) $f^{-1}(x) = f(y) = f(y_0) + \frac{\partial f}{\partial y}_{(y_0)}(y - y_0) + \frac{1}{2}\frac{\partial^2 f}{\partial y \partial y}_{(y_0)}(y - y_0)^2 + \frac{1}{6}\frac{\partial^3 f}{\partial y \partial y \partial y}_{(y_0)}(y - y_0)^3 = e + e(y - e) = e + ey - e^2$. Particularly, $f(e) = e$.

3.6. CASE STUDY: MINIMAL SURFACE AREA

Case 3.1. Let us take an open box with no lid, as shown (Fig. **3.6**). The box has volume 47 and dimensions x_1, y_1, z_1. Using the constraint to substitute for z_1, find the dimensions of the box with a minimal surface area. **Case adapted with permission of the author** [5]. (i) What is the surface area of the box in terms of just x_1 and y_1? (ii) Using (i), find the values of x_1, y_1, z_1 that minimize the surface area of the box.

Solution 3.13. (i) The surface area is to be expressed in terms of x and y_1.

In order to have an equation for the surface area without z_1, we start with a direct expression for it

$$\text{Area} = 2x_1y_1 + 2x_1z_1 + y_1z_1$$

Then, we get the volume constraint and solve for z_1 in terms of x_1, y_1

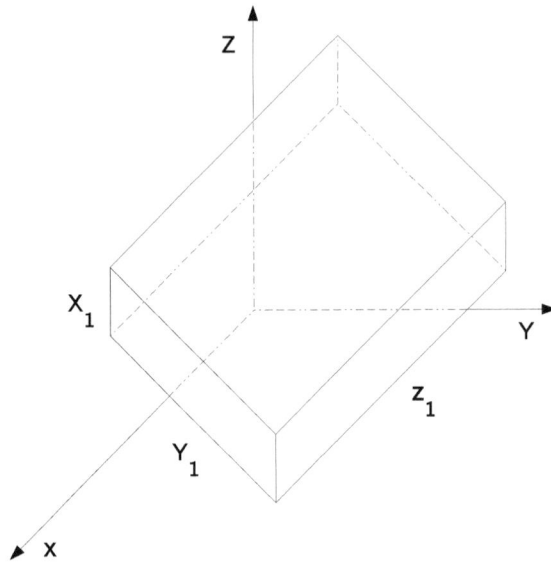

Fig. (3.6). Surface area of the box.

$$x_1 y_1 z_1 = 47$$

$$z = \frac{47}{x_1 y_1} \tag{3.3}$$

Substituting z_1 (Eq. 3.3) in the expression for the area, we get

$$A(x, y) = 2x_1 y_1 + \frac{94}{y_1} + \frac{47}{x_1}. \tag{3.4}$$

(ii) Finding the minimal surface area. This is an unconstrained minimization problem in x and y. We know that because these variables can be any positive number, the constraint of the volume is 47 as stated in the expression $A(x, y)$ (Eq. 3.4).

Recall:

To find the edge points of a function without constraint, we have to find the critical points by setting all partial derivatives equal to zero, then we compute the partial derivatives of $A(x, y)$:

$$\partial_x A(x, y) = 2y_1 - \frac{47}{x_1^2} \tag{3.5}$$

$$\partial_y A(x,y) = 2x_1 - \frac{94}{y_1^2} \tag{3.6}$$

Setting these derivatives to 0, we see a critical point can only be located at (x,y) satisfying that

$$2y_1 - \frac{47}{x_1^2} = 0 \tag{3.7}$$

$$2x_1 - \frac{94}{y_1^2} = 0 \tag{3.8}$$

Solving for y (Eq. 3.7) and plugging in (Eq. 3.8), we get $x_1 - 0.085x_1^4 = 0$. To solve for x_1 we factor

$$x_1 - 0.085x_1^4 = 0.$$

This expression is zero if $x_1 = 0$ or $1 - 0.085x_1^3 = 0$. We ignore the $x_1 = 0$ solution because A is undefined for $x_1 = 0$. Hence, the only critical point is at $x_1 = 2.27$. We can now use (Eq. 3.3) and (Eq. 3.8) to find that $y_1 = 4.55$ and $z_1 = \frac{47}{2.27 \times 4.55} = 4.55$.

We suppose that the minimal surface area is achieved with

$$x_1 = 2.27, y_1 = 4.55, z_1 = 4.55.$$

We could use the second derivative test to verify that these values indeed provide a minimum.

3.7. CASE STUDY: MAXIMUM SALES VOLUME

Case 3.2. The volume of sales of a product is a function of the number of ads in newspapers x and the number of minutes on television y. (**Case adapted with permission of the author** [6].

Statistically, these variables are related this way

$$V = 12xy - x^2 - 3y^2.$$

An ad in the press costs 60 euros and a minute on television 600 euros. The advertising budget is 9000 euros. Determine the optimal advertising policy.

The function to optimize is $V(x,y) = 12xy - x^2 - 3y^2$, based on $-60x - 600y = 9000$. Be $g(x,y) = -60x - 600y - 9000$. Then $\nabla V(x,y) = \lambda g(x,y)$. So the system to solve is

$$-2x + 12y + 60\lambda = 0 \qquad\qquad (3.9)$$

$$12x - 6y + 600\lambda = 0$$

$$60x + 600y - 9000 = 0$$

Whose critical point is $(\lambda, x, y) = (-\frac{165}{223}, \frac{9450}{223}, \frac{2400}{223})$

Now, we evaluate this critical point in (Eq. 3.2), bearing in mind that $h = V - \lambda g$

$$H_3\big(f(x)\big) = \begin{pmatrix} 0 & \dfrac{\partial g}{\partial x} & \dfrac{\partial g}{\partial y} \\[2mm] \dfrac{\partial g}{\partial x} & \dfrac{\partial^2 h}{\partial x\,\partial x} & \dfrac{\partial^2 h}{\partial x\,\partial y} \\[2mm] \dfrac{\partial g}{\partial y} & \dfrac{\partial^2 h}{\partial y\,\partial x} & \dfrac{\partial^2 h}{\partial y\,\partial y} \end{pmatrix}_{x_0}$$

$$= \begin{pmatrix} 0 & -60 & -600 \\ -60 & -2 & 12 \\ -600 & 12 & -6 \end{pmatrix}_{\left(-\frac{165}{223}, \frac{9450}{223}, \frac{2400}{223}\right)}$$

$$H_3\big(f(x)\big) = \begin{bmatrix} 0 & -60 & -600 \\ -60 & -2 & 12 \\ -600 & 12 & -6 \end{bmatrix} = 1065600 > 0$$

Thus, the critical point $(\lambda, x, y) = \left(-\frac{165}{223}, \frac{9450}{223}, \frac{2400}{223}\right)$ is a maximum point. Therefore, the optimal advertising policy will be getting $\frac{9450}{223} \approx 42$ ads in the press and $\frac{2400}{223} \approx 10$ ads on tv.

3.8. EXERCISES

Exercise 3.1. Show that $T(x, t) = e^{-kt}\cos x$ ([4], page 157) satisfies $k\frac{\partial^2 T}{\partial x\, \partial x} = \frac{\partial T}{\partial t}$.

Exercise 3.2. Find the critical points of [35] $f: \mathbb{R}^2 \to \mathbb{R}, x^2 + y^2 + x + y + xy$.

Exercise 3.3. Find the critical points of [35] $f(x, y) = \log(1 + x^2 + y^2) - \int_0^x \frac{2t}{1+t^4}\, dt$.

Exercise 3.4. Find the critical points of [36] $f: \mathbb{R}^2 \to \mathbb{R}, x^3 - 6xy + y^3$.

Exercise 3.5. Find the critical points of the plane $f: \mathbb{R}^2 \to \mathbb{R}, 1$.

Exercise 3.6. Let the function $f(x, y) = x^3 + y^3$. (i) Determine its critical points. (ii) What kind of critical points are they?

Exercise 3.7. Let the real-valued function $f: S \subset \mathbb{R}^2 \to \mathbb{R}, x + y$, where $S = \{(x, y) \in \mathbb{R}^2 |\ x^2 + y^2 = 1,\ x, y \in \mathbb{R}\}$. Determine its critical points on the **boundary** of S

Exercise 3.8. Let the function $x^2 + y^2 + z^2 - 9 = 0$. (i) Is it possible to define a function f, with respect to the variable x, in the neighbourhood of point $(1,2,2)$? (ii) What is the total derivative of function f? (iii) What is the polynomial approximation of function f in order 2?

Exercise 3.9. Let the function $f(x) = \sin x$. (i) Is it possible to define a function f^{-1}, with respect to the variable y, in the neighbourhood of point $x = 0$? (ii) What is its derivative? (iii) What is the function (or polynomial approximation) of f^{-1} in order 3?

CHAPTER 4

Vector-Valued Functions

Abstract: This chapter focuses on describing the vector-valued function F, with the graphs called **vector fields**, and the characterisation of this vector field with three operators: **lines of flow**, **rotational rot**, and **divergence div**.

Keywords: Class C^1 function, Conservative vector field, Counter-clockwise circle, Divergence, Gauss' theorem, Graph of vector valued-function, Gravitational field, Green's theorem, Lines of flow, Paddle wheel, Real valued-function, Rotational, Stokes' theorem, Vector fields, Vector-valued function.

4.1. VECTOR-VALUED FUNCTIONS

Vector-valued functions can be applied to all scientific and technological fields. They are a useful tool to abstract mathematically, and sometimes graphically, the behaviour of all "entities" that have magnitude and direction. The vectors generated in a system can give a general but also a particular outlook on the behaviour of a vector' valued function; this graphic representation is called "flow field".

Rotational and divergence operators will also be described. These operators measure, locally or in a group, the convergence or rotation of a vector field. They will be explained in two sections: here, as a metric describing the local movement of a vector field and later as part of the integrals of Green's , Stokes', and Gauss' theorems (Ch 8) that measure the global flow of a vector field.

Definition 4.1. A vector-valued function is a transformation $F : U \subset \mathbb{R}^n \to \mathbb{R}^n$ that assigns to each point $x \in U$ a vector $F(x)$. The vector-valued function $F(x, y) = F_1(x, y), F_2(x, y))$. The real-valued functions F_1 and F_2 are the components of the vector-valued function F. The **graph** of function F is called **vector field**.

Note 4.1. Vector-valued functions F are very useful to associate physical quantities with a specific position.

The procedure for plotting a vector (x_0, y_0) in the Cartesian coordinate system is:

Role 1. Plot the point (x_0, y_0) in the Cartesian coordinate system.

Role 2. Evaluate the point (x_0, y_0) in the vector-valued function $F(x, y)$.

Role 3. Move the Cartesian coordinate system to the point (x_0, y_0).

Role 4. Plot the vector $F(x_0, y_0)$ in the new Cartesian coordinate system.

Example 4.1. Let the function $F(x, y) = (x, 2y)$. (i) What is the vector associated with point $(1,1)$? (ii) What is its vector field?

Solution 4.1. (i) The vector c associated with point $(1,1)$ is $(1,2)$. (ii) Its vector field is the representation of vector $c = (1,2)$ in the Cartesian coordinate system (Fig. **4.1**).

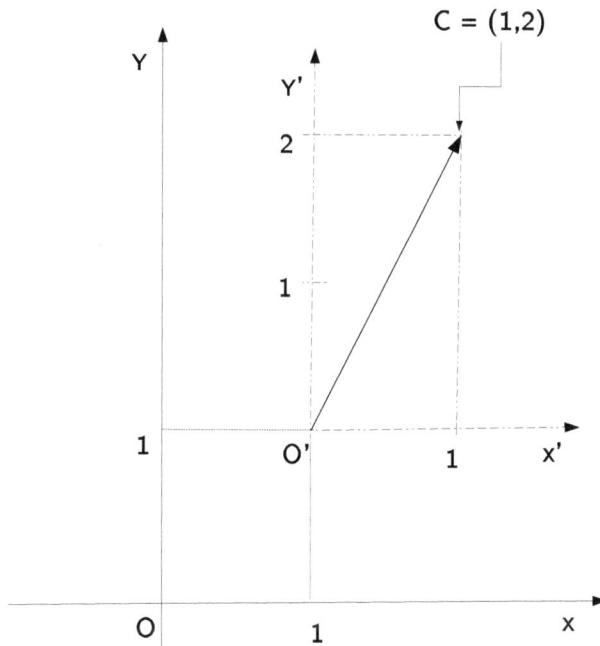

Fig. (4.1). The point $(1,1) \in \mathbb{R}^2$ is associated with the **vector-valued function** $F(x, y) = (x, 2y)$ with the vector $(1,2)$.

Example 4.2. Let the function $F(x, y) = -yi + xj = (-y, x)$. What is its vector field?

Solution 4.2. Its **vector field** is shown in Fig. (**4.2**).

Graph of the vector-valued function

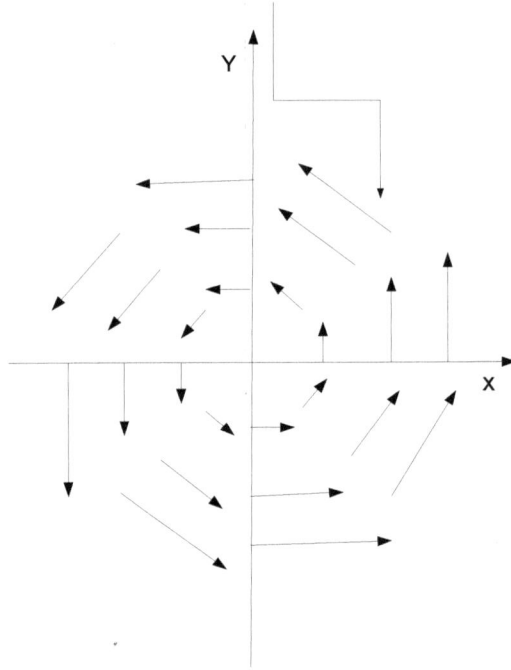

Fig. (4.2). Graph of the **vector-valued function** $F(x, y) = (-y, x)$, *i.e.* the vector field.

4.1.1. Graph of a Vector-Valued Function

Definition 4.2. A vector field is a vector-valued function $F: \mathbb{R}^n \to \mathbb{R}^n$ that assigns to each point of its domain a vector that acts on that point.

Note 4.2. Examples of vector fields are the functions that represent electric fields, gravitational fields, the movement of the wind next to an aerodynamic surface, ocean currents, and the speed of a fluid.

Their graphical representations are the graphs, level surfaces, and flow lines.

Example 4.3. The **gravitational field** is the influence or interaction that an object of mass m experiences with respect to another of mass M, subject to $m \ll M$, where r is the distance between the centres of the spheres that cover the objects and the gravitational force $F(x, y, z) = -\frac{MmG}{r^2}$.

$$g(x, y, z) = \frac{F}{m} = -\frac{(\frac{MmG}{r^2}x, \frac{MmG}{r^2}y, \frac{MmG}{r^2}z)}{m} = -(\frac{MG}{r^2}x, \frac{MG}{r^2} \text{ and}, \frac{MG}{r^2}z).$$

What is the acceleration of the gravitational field that affects a subject on the surface of the Earth? (Fig. **4.3**).

Solution 4.1. Let $G = 6.67 \times 10^{-11} \frac{N\,m^2}{kg^2}$. The mass of the subject is $m = 70kg$. The mass of planet Earth is $M = 5.98 \times 10^{24} kg$. The radius of the Earth is $r = 6.38 \times 10^6 m$. Then, the acceleration of the gravitational field on the surface is $g(x, y, z) = -9.80m/s^2$.

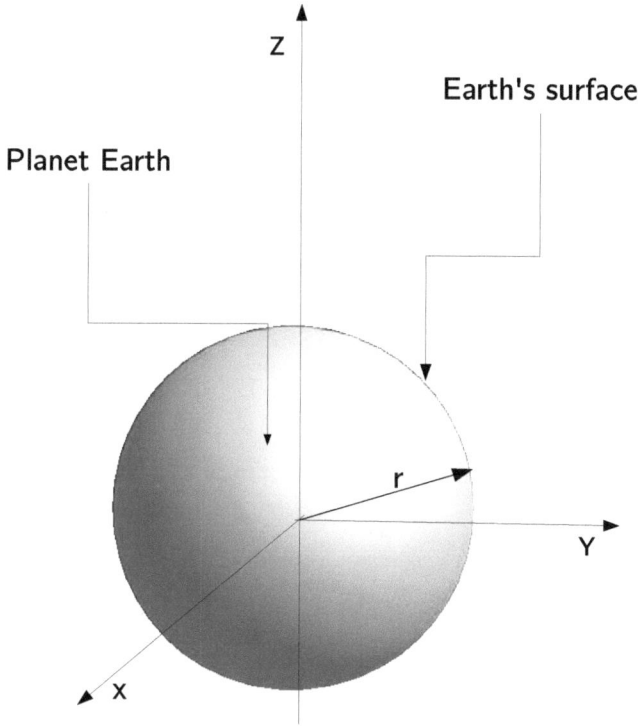

Fig. (4.3.). The gravitational field points to the interior of planet Earth.

4.1.2. Flow Line of a Vector Field

A **flow line** $c: \mathbb{R} \to \mathbb{R}^n$ (Eq. 4.1) is a curve that describes the direction or tendency of the vectors that form a **vector field** and analytically satisfies

$$\frac{\partial c}{\partial t} = F \circ C(t). \tag{4.1}$$

Example 4.4. Let the vector field $F(x, y) = (-y, x)$ (Fig. **4.2**). (i) Describe the general tendency of this vector field. (ii) What is the flow line?

Solution 4.2. (i) The tendency of this vector field is a circular curve with counter-clockwise orientation (Fig. **4.4**). (ii) Suppose curve $C(t) = (\cos t, \sin t), t \in [0, 2\pi]$, this curve meets $C'(t) = (-\sin t, \cos t) = F \circ C(t) = F(C(t)) = F(\cos t, \sin t) = (-\sin t, \cos t)$. Then, the curve $C(t)$ is the flow line.

Note 4.3. In fact, any concentric curve with that orientation is the flow line of this vector-valued function.

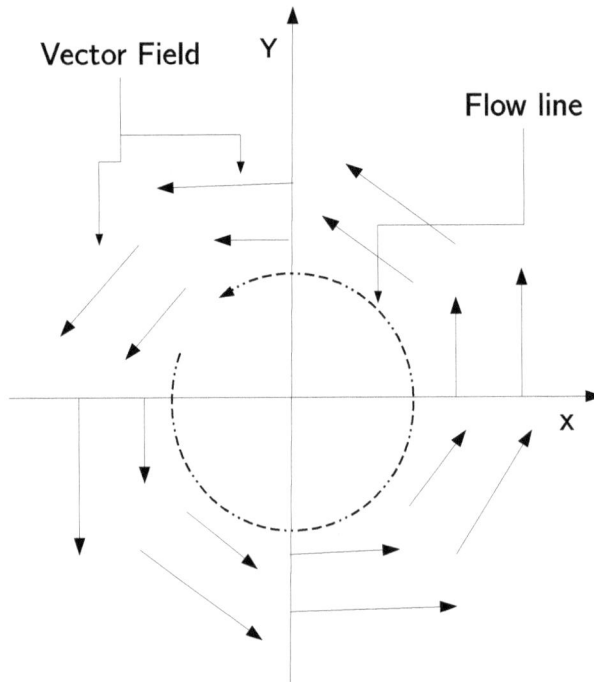

Vector-valued function F(x, y) = (-y, x)

Fig. (4.4). The **flow line** of the vector field is a circular curve with counter-clockwise orientation.

4.1.3. Rotational of a Function

The **rotational** operator *rot F* of a vector-valued function F is the measure of the rotation of the flow, *i.e.* the vectors that form a vector field around a particular point conforming a tiny paddle wheel at the point evaluated (Fig. **4.5**). There is a difference in the intensity of the vector field as it is represented by vectors of different lengths above and below the paddle wheel, thus inducing a **rotational movement** around the point.

Note 4.4. The physical sense of this operator will be treated later (Ch. 8) because it requires the use of vector integrals.

Definition 4.3. For a vector-valued function $F: \mathbb{R}^3 \to \mathbb{R}, (F_1(x), F_2(x), F_3(x))$, where real-valued functions $F_i: \mathbb{R}^n \to \mathbb{R}, i = 1, \cdots, 3$ are of class C^1. The rotational (Eq. (4.2)) of vector-valued function F at point x_0 is the **vector-valued function** defined by $rot\ F$.

$$rot\ F_{x_0} = \begin{pmatrix} i & j & k \\ \frac{\partial}{\partial x} & \frac{\partial}{\partial y} & \frac{\partial}{\partial z} \\ F_1 & F_2 & F_3 \end{pmatrix}_{x_0} = \left(\frac{\partial F_3}{\partial y} - \frac{\partial F_2}{\partial z}\right) i - \left(\frac{\partial F_3}{\partial x} - \frac{\partial F_1}{\partial z}\right) j + \left(\frac{\partial F_2}{\partial x} - \frac{\partial F_1}{\partial y}\right) k \qquad \textbf{(4.2)}$$

The physical interpretation of $rotF$ is an infinitely small paddle wheel at point (x_0, y_0) inside the vector field.

Example 4.5. Let the vector-valued function $F: \mathbb{R}^3 \to \mathbb{R}, x^3y + x^2z^2 - \sin xyz$. (i) What is the $rot\ F$? (ii) What is the $rot\ F(0,1,2)$? (iii) Is the $rot\ F(0,1,2)$ truly representative of the rotational of F on the open set $\mathbb{R}^3$?

Solution 4.3.

$$(i)\ rot\ F_{x_0} = \begin{pmatrix} i & j & k \\ \frac{\partial}{\partial x} & \frac{\partial}{\partial y} & \frac{\partial}{\partial z} \\ x^3y & x^2z^2 & -\sin xyz \end{pmatrix}_{x_0}$$
$$= (-xz\cos xyz - 2x^2z, yz\cos xyz, 2xz^2 - x^3)$$

(ii) $rot\ F(0,1,2) = (-xz\cos xyz - 2x^2z, yz\cos xyz, 2xz^2 - x^3)_{(0,1,2)} = 2.$

(iii) No, it is not. The $rot\ F$ operator is a **local** operator, so its value will depend on each point.

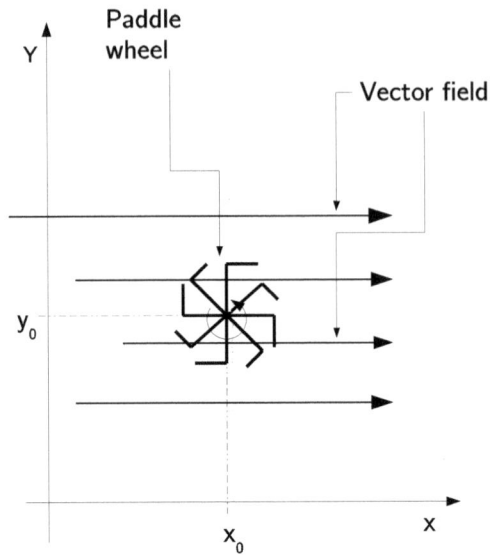

Fig. (4.5). The physical interpretation of rot F is an infinitely small paddle wheel at point (x_0, y_0) inside the vector field.

4.1.4. Divergence of a Function

The divergence operator *div F* of a vector-valued function *F* is a measure of the expansion or contraction of the vector field per unit of volume or area, around the point x_0 (Fig. **4.6**).

Note 4.5. The physical sense of this operator will be treated later (Ch. 8) because it requires the use of vector integrals.

Definition 4.4. For a vector-valued function $F: \mathbb{R}^n \to \mathbb{R}, (F_1(x), F_2(x), \cdots, F_n(x))$, where the real-valued functions $F_i: \mathbb{R}^n \to \mathbb{R}, i = 1, \cdots, n$, are of class C^1, the divergence (Eq. 4.3) of vector-valued function F at point x_0 is the **real-valued function** defined by

$$div\ F_{x_0} = \nabla \cdot F_{x_0} = \frac{\partial F_1}{\partial x_1} + \frac{\partial F_2}{\partial x_2} + \cdots + \frac{\partial F_n}{\partial x_n} \qquad (4.3)$$

Example 4.6. Let the vector-valued function $F: \mathbb{R}^3 \to \mathbb{R}, x^3y + x^2z^2 - \sin xyz$. (i) What is the *div F*? (ii) What is the *div F*(0,1,2)? (iii) Is the *div F*(0,1,2) truly representative of the divergence of F on the open set $\mathbb{R}^3$?

(i) The divergence of function F is $div\ F = \nabla \cdot F = \frac{\partial F_1}{\partial x_1} + \frac{\partial F_2}{\partial x_2} + \frac{\partial F_3}{\partial x_3} = 3x^2y + 0 -$ $xy\cos xyz$. (ii) $div\ F(0,1,2) = 0$. (iii) No, it is not. The $div\ F$ operator is a **local** operator, so its value will depend on each point.

Volume density around the point

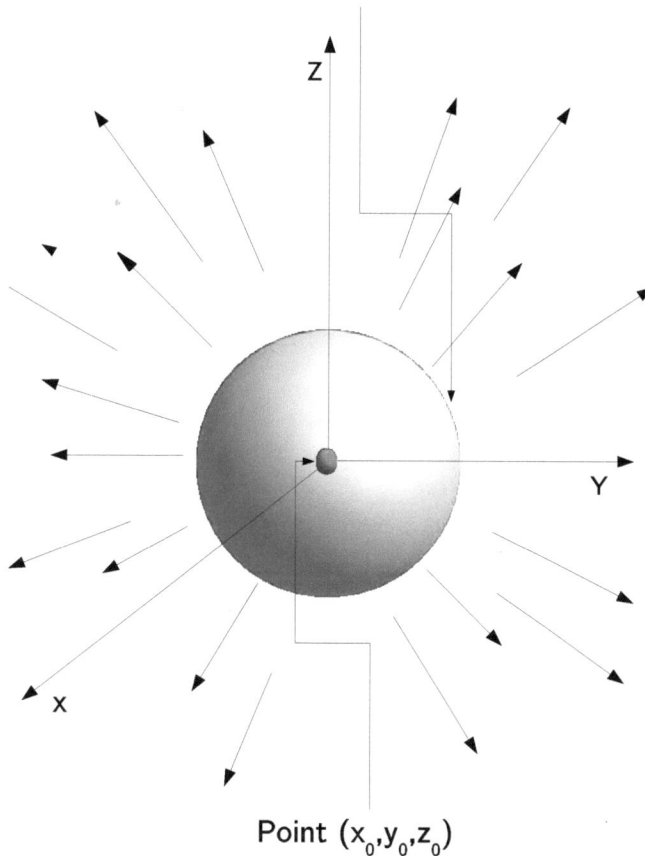

Point (x_0, y_0, z_0)

Fig. (4.6). $div\ F > 0$ so the vector field seems to explode from the origin.

4.2. CASE STUDY: POTENTIAL FUNCTION OF A VECTOR FIELD

Case 4.1. Is $\mathbf{F}(x,y) = \frac{1}{(x+y)^2}\mathbf{i} + \frac{1}{(x+y)^2}\mathbf{j}$ conservative? If so, find a $\phi(x,y)$ such that $\mathbf{F}(x,y) = \nabla\phi(x,y)$ **Case adapted with permission of the author** [7]. (i) Determining if $F(x,y)$ is conservative. The vector field $\mathbf{F}(x,y) = u(x,y)\mathbf{i} + v(x,y)\mathbf{j}$ is conservative if, and only if,

$$\partial_y u(x,y) = \partial_x v(x,y).$$

We identify $u(x,y) = \frac{1}{(x+y)^2}$ and $v(x,y) = \frac{1}{(x+y)^2}$, then we can compute the partial derivatives

$$\partial_y u(x,y) = -\frac{2}{(x+y)^3} \tag{4.4}$$

$$\partial_x v(x,y) = -\frac{2}{(x+y)^3} \tag{4.5}$$

They are equal, therefore, $F(x,y)$ is conservative.

(ii) Finding the potential $\phi(x,y)$. Let us recall the definition of a conservative vector field.

A vector field $F(x,y)$ is conservative if there exists a $\phi(x,y)$ such that $F(x,y) = \nabla\phi(x,y)$.

Futhermore, the function $\phi(x,y)$ can be determined by integrating the equations of each component of

$$\mathbf{F}(x,y) = \nabla\phi(x,y) = \partial_x\phi(x,y)\,\mathbf{i} + \partial_y\phi(x,y)\,\mathbf{j}$$

and combining the results into a single function ϕ.

Expressing the components

$$\frac{1}{(x+y)^2} = \partial_x\phi(x,y) \tag{4.6}$$

$$\frac{1}{(x+y)^2} = \partial_y\phi(x,y) \tag{4.7}$$

Integrating (Eq. 4.6) with respect to x and (Eq. 4.7) with respect to y gives

$$\phi(x,y) = -\frac{1}{x+y} + g(y) \tag{4.8}$$

$$\phi(x,y) = -\frac{1}{x+y} + h(x) \tag{4.9}$$

For any $g(y)$ and $h(x)$

These relations are both satisfied by

$$\phi(x, y) = \frac{1}{x+y} + c$$

for any constant c.

4.3. CASE STUDY: $F_a(x, y) = \left(\frac{-y}{r^a}, \frac{x}{r^a}\right)$ On The Plane

Case 4.2. For any number $a \in \mathbb{R}$ we have $F_a(x, y) = (F_1, F_2) = (\frac{-y}{r^a}, \frac{x}{r^a})$, where $r = \sqrt{x^2 + y^2}$. **Case reproduced with permission of the author** [8]. Note that all of these vector fields differ from the $a = 0$ case just by rescaling, particularly because all the trajectories are circles. In the last section, we saw that when $a = 0$ the rotational was equal to 2. Let us see what happens in general when the scalar part of the rotational is $\frac{\partial F_2}{\partial x} - \frac{\partial F_1}{\partial y}$.

We compute $\frac{\partial r}{\partial x}$ and $\frac{\partial r}{\partial y}$ in advance.

We have

$$\frac{\partial r}{\partial x} = \frac{2x}{2\sqrt{x^2+y^2}} = \frac{x}{r},$$

$$\frac{\partial r}{\partial y} = \frac{y}{r}.$$

$$\frac{\partial F_2}{\partial x} = \frac{d(\frac{x}{r^a})}{dx}\frac{1}{r^a} + x(\frac{d\frac{x}{r^a}}{dx}) = r^{-a} + x(-a)r^{-a-1}\frac{dr}{dx}$$

$$= r^{-a} - a(x)r^{-a-1}\frac{x}{r} = r^{-a} - ar^2 r^{-a-2} = \frac{r^2 - ax^2}{r^{a+2}}.$$

and

$$\frac{\partial F_2}{\partial y} = -\frac{d(-\frac{y}{r^a})}{dy} = \frac{d(\frac{y}{r^a})}{dy} = \frac{r^2 - 2ax^2}{r^{a+1}}.$$

Thus, the scalar of the rotational is

$$\frac{2(r^2 - a(x^2+y^2))}{r^{a+2}} = \frac{(2-a)(x^2+y^2)}{r^{a+2}}.$$

When $a < 2$ the rotational is positive, particularly in the case $a = 0$, and when $a >$ 2 the rotational is negative. In this case, despite the fact that any given particle travels around in a counter-clockwise circle, a paddle wheel nailed to any point will spin clockwise. Furthermore, at the special case $a = 2$ we get an irrotational vector field,

$$F_{(x,y)} = \left(\frac{-y}{x^2+y^2}, \frac{x}{x^2+y^2}\right).$$

4.4. EXERCISES

Exercise 4.1. Draw the vector-valued function $F(x,y) = (y,x)$.

Exercise 4.2. Let the vector-valued function $F(x,y,z) = (x,y,z)$. (i) What is its divergence at point $(2,3,1)$. (ii) What is its rotational at point $(7,2,1)$. (iii) What is the flow line at point $(1,1,1)$?

Exercise 4.3. Show div rot $F = 0$, where $F: \mathbb{R}^3 \to \mathbb{R}, (F_1(x), F_2(x), F_3(x))$ and the real-valued functions $F_i: \mathbb{R}^n \to \mathbb{R}, i = 1 \cdots 3$ are class C^1.

Exercise 4.4. Let the vector-valued function $F(x,y) = (-y,x)$. (i) What is its divergence at point $(2,3)$. (ii) What is its rotational at point $(2,1)$. (iii) What is the flow line at point $(1,1)$?

Exercise 4.5. A vector field is called a **conservative vector field** [37] if there exists a function such that $F = \nabla f$. If F is a conservative vector field, then the function f is called a **potential function** for F. Provide an example of conservative vector field and potential function.

Exercise 4.6. Determine the equation of flow lines [38] or field lines of $F(x,y) = (1,y)$.

Exercise 4.7. Define the next operations as Vector, Scalar, or Nonsense [39]: (i) $rot(\nabla f)$; (ii) div$(rot f)$; (iii) $\nabla \cdot (\nabla \times F)$.

Part III

INTEGRATION

CHAPTER 5

Integration Over Bounded Regions

Abstract: This chapter introduces Riemann's double and triple integrals over integration domains, bounded by real-valued functions with and without the use of mappings that transform the bounded integration domains.

Keywords: Bounded regions, Bounded and not closed region, Closed and bounded regions, Double Riemann integral, Integration domains, Jacobian determinant, Lower bound region, Maps in $\mathbb{R}$, Maps in $\mathbb{R}^2$, Maps in $\mathbb{R}^3$, Not closed and not bounded region, Simple Riemann integral, Triple Riemann integral, Upper bound region.

5.1. BOUNDED REGIONS

Definition 5.1. A bounded region is a region that has an upper and a lower bound. An unbounded region has the opposite characteristics, the upper and/or lower bounds are not finite [40].

Example 5.1. Provide examples of closed and bounded sets.

Solution 5.1. Closed and bounded: $[0,2]$. Closed and not bounded: $\cup_{n\in Z} [3n, 3n + 1]$. Bounded and not closed: $(0,2)$. Not closed and not bounded: $\cup_{n\in Z} (3n, 3n + 1)$ [41].

5.2. MAPS $\mathbb{R} \to \mathbb{R}$

Definition 5.2. Let D and D^* be affine spaces in $\mathbb{R}$ $T: D^* \to D$ (Fig. **5.1**) of the form $x \mapsto Mx + v$, where M is a linear transformation on D^* and v is a vector in D [42].

Example 5.2. Consider the map $T: [0,2] \subset \mathbb{R} \to [0,4] \subset \mathbb{R}, (2x)$. (i) What is the domain of T? (ii) What is the image of T?

Solution 5.2. (i) The domain of T is the closed interval $[0,2]$. (ii) The image of T is the closed interval $[0,4]$.

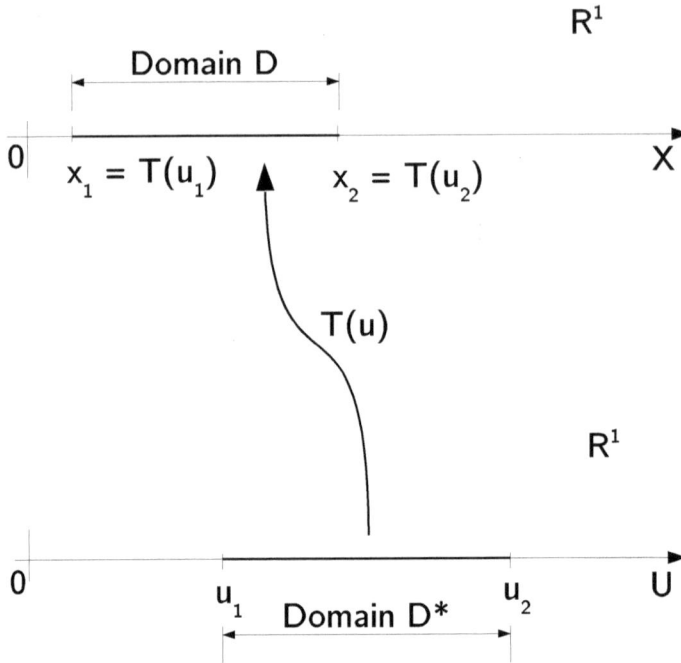

Fig. (5.1). Map $T: [u_1, u_2] \subset D^* \to [x_1, x_2] \subset D$, where $x = T(u)$.

5.3. JACOBIAN DETERMINANT ON $\mathbb{R}$

Definition 5.3. Let T be a transformation such that $T: u \in \mathbb{R} \to x \in \mathbb{R}$, the Jacobian determinant (Eq. 5.1) or Jacobian $J(T)$ is the determinant of the *matrix derivatives* of T (Thm. 3.2).

$$J(T(u)) = \left| \frac{dT}{du} \right| \tag{5.1}$$

Example 5.3. Consider the map $T: \mathbb{R} \to \mathbb{R}, 2u$. What it the Jacobian determinant?

Solution 5.3.

$$J(T(u)) = \left| \frac{dT}{du} \right| = 2 \tag{5.2}$$

5.4. SIMPLE RIEMANN INTEGRAL

In this section, we present the resolution of a simple Riemann integral using an $\mathbb{R}$ in $\mathbb{R}$ mapping, which affects the integration domain and the function to integrate. This simplifies both elements getting an equivalent solution.

Definition 5.4. The simple Riemann integral (Eq. 5.2) of a real-valued function f over an interval $D = [a, b]$, where $[\overline{x}_i, \overline{x}_j]$ are the small sub-intervals of D and $\overline{x}_i$, $\overline{y}_j$ are the midpoints of these intervals, is

$$\int_D f(x)dx = \lim_{i \to \infty} \sum_{i=1}^n f(\overline{x}_i) \, dx \qquad (5.2)$$

Definition 5.5. For a real-valued function $f: D \subset \mathbb{R} \to M \subset \mathbb{R}$ and a map $T: D^* \subset \mathbb{R} \to D \subset \mathbb{R}$, the simple Riemann integral is equivalent to (Eq. 5.3)

$$\begin{aligned}
\int_D f(x) \, dx &= \int_{x_1}^{x_2} f(x) \, dx \\
&= \int_{D^*} f(x) \circ T(u) \, |J(T(u)| \, du \qquad (5.3) \\
&= \int_{u_1}^{u_2} f(T(u)) \, |J(T(u))| \, du.
\end{aligned}$$

Example 5.4. Let the function $f(x) = x^2\sqrt{x^3}, x \in [1,2]$. (i) What is the area under the curve of that interval? (ii) What map is convenient to simplify this integral? (iii) Solve the integral using the map $T(u)$. (iv)Solve the integral using the change of variable method $u = x^3$. (v) Do the geometric analysis of both graphs.

Solution 5.4. (i) $\int_1^2 x^2\sqrt{x^3} \, dx = \left[\frac{2}{9}x^{\frac{9}{2}}\right]_1^2 = 4.8060$. (ii) Let $T(u) = u^{\frac{1}{3}}, u \in [1,8]$.

(iii) If $T(u) = u^{\frac{1}{3}}$, then $J(T(u)) = \frac{1}{3u^{\frac{2}{3}}}$, $\int_1^2 x^2\sqrt{x^3} \, dx = \int_1^8 f \circ T(u)|J(T(u)|du =$

$\frac{1}{3} \int_1^8 u^{\frac{2}{3}}\sqrt{u} \, u^{-\frac{2}{3}} \, du = \frac{1}{3}\int_1^8 \sqrt{u} \, du = 4.8061$. (iv) If $u = x^3 \Rightarrow du = 3x^2 \, dx \Rightarrow$

$dx = \frac{du}{3x^2} \Rightarrow \int_1^8 x^2\sqrt{u} \, \frac{du}{3x^2} = \frac{1}{3}\int_1^8 \sqrt{u} \, du = 4.8061$. (v) The closed areas of both functions $f_1(x) = x^2\sqrt{x^3}$ and $f_2(x) = \frac{1}{3}\sqrt{x^3}$, on different integration domains are equivalent (Fig. **5.2**).

Example 5.5. Let the function $f(x) = \sqrt{1 - x^2}, x \in [-1,1]$. (i) What is the area under the curve of that interval? (ii) What map is convenient to simplify this integral? (iii) Solve the integral using the map $T(\theta)$. (iv) Solve the integral using the change of variable method $u = \arcsin\theta$. (v) Do the geometric analysis of both graphs.

Solution 5.5. (i) $\int_{-1}^1 \sqrt{1 - x^2} \, dx = \frac{1}{2}\left[x\sqrt{1 - x^2} + \sin x\right]_{-1}^1 = \frac{\pi}{2}$. (ii) $T(\theta) = \sin\theta$, $\theta \in [-\frac{\pi}{2}, \frac{\pi}{2}]$. (iii) Let $T(\theta) = \sin\theta, J(T(\theta)) = \cos\theta$, then $\int_{\theta_1}^{\theta_2} f \circ T(\theta) \, |J(T(\theta)|d\theta$

$$= \int_{-\frac{\pi}{2}}^{\frac{\pi}{2}} \cos\theta \, \cos\theta \, d\theta = \frac{\pi}{2}. \text{ (iv) If } x = \sin\theta \Rightarrow dx = \cos\theta \, d\theta \int_{\arcsin(-1)}^{\arcsin(1)} \cos^2\theta \, d\theta$$

$$= \int_{-\frac{\pi}{2}}^{\frac{\pi}{2}} \cos^2\theta \, d\theta = \frac{\pi}{2}. \text{ (v) The closed areas of both functions } f_1(x) = \sqrt{1-x^2} \text{ and}$$

$f_2(x) = \cos^2 x$ on different integration domains are equivalent (Fig. **5.3**).

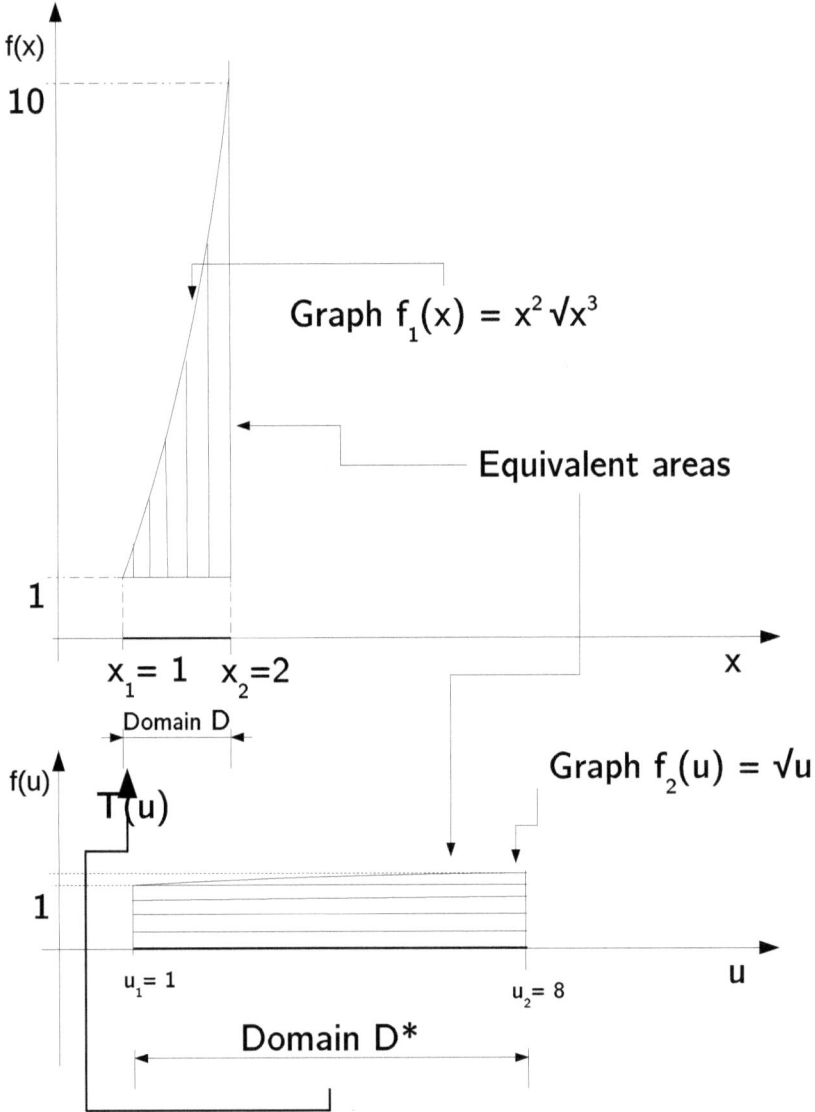

Fig. (5.2). The T mapping goes from D^* space to D space. The area under $f_1(x) = x^2\sqrt{x^3}$ over [1,2] and the area under $f_2(u) = \sqrt{u^3}$ over [1,8] are equivalent.

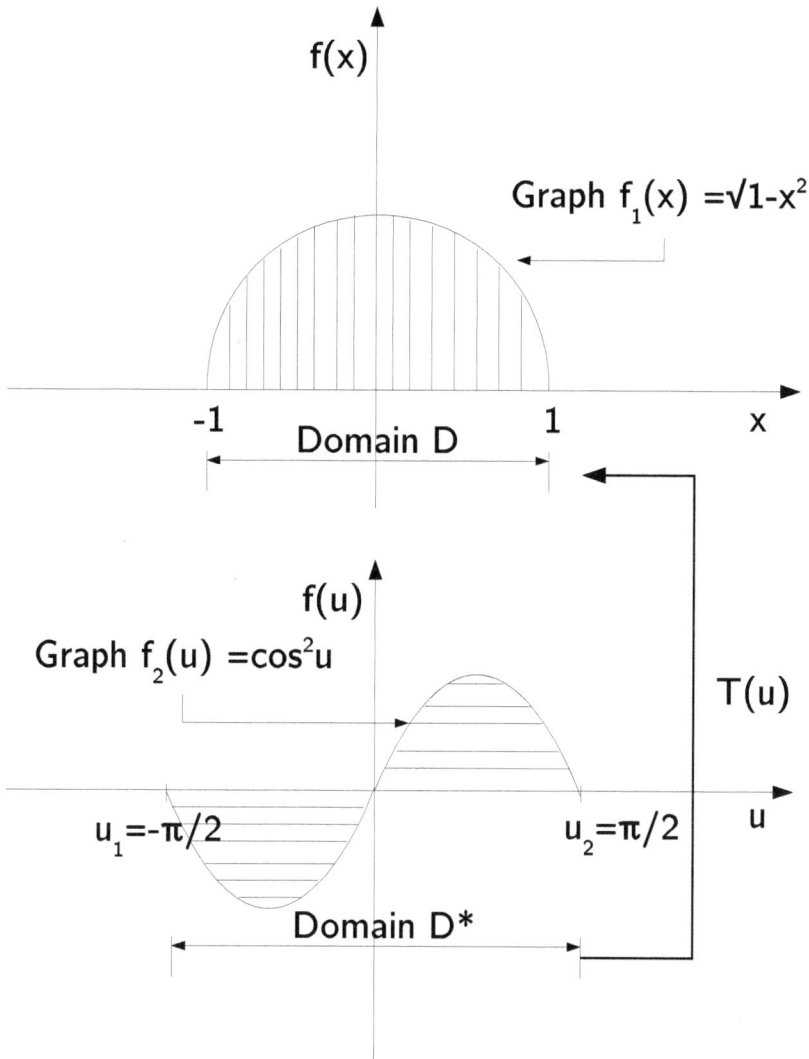

Fig. (5.3). The T mapping goes from D^* space to D space. The area under $f_1(x) = \sqrt{1-x^2}$ over $[-1,1]$ and the area under $f_2(x) = \cos^2 x$ over $[-\frac{\pi}{2},\frac{\pi}{2}]$ are equivalent.

5.5. MAPS $\mathbb{R}^2 \to \mathbb{R}^2$

Definition 5.6. Let D and D^* be affine spaces in $\mathbb{R}^2$ $T\colon D^* \to D$ (Fig. **5.4**) of the form $x \mapsto Mx + v$, where M is a linear transformation on D^* and v is a vector in D [42].

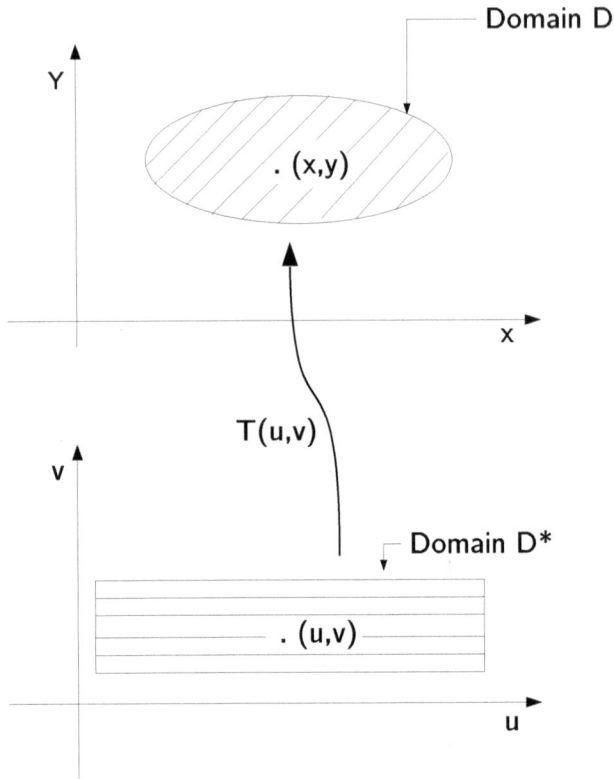

Fig. (5.4). Map $T: D^* \subset \mathbb{R}^2 \to D \subset \mathbb{R}^2$, where $T(D^*) = D$.

Example 5.6. Let the map $T: [0,1] \times [0,2\pi] \subset \mathbb{R}^2 \to \mathbb{R}^2, (r\cos\theta, r\sin\theta)$. (i) What is the domain of T? (ii) What is the image of T?

Solution 5.6. (i) The domain of T on the $r\theta$-plane is the rectangle $[0,1] \times [0,2\pi]$. (ii) The image of T is a unit circle with the centre at the origin of the xy-plane.

5.6. JACOBIAN DETERMINANT ON $\mathbb{R}^2$

If T is a transformation such that $T: (u, v) \in \mathbb{R}^2 \to (x, y) \in \mathbb{R}^2$, the Jacobian determinant (Eq. 5.4) or Jacobian $J(T)$ is the determinant of *the matrix derivatives* of T (Thm. 3.2)

$$J(T(u,v)) = \begin{vmatrix} \dfrac{dT_1}{du} & \dfrac{dT_1}{dv} \\[2mm] \dfrac{dT_2}{du} & \dfrac{dT_2}{dv} \end{vmatrix} \tag{5.4}$$

Example 5.7. Let T be a transformation such that $T: \mathbb{R}^2 \to \mathbb{R}^2, (r\cos\theta, r\sin\theta)$. What is the Jacobian determinant?

Solution 5.7.

$$J(T(u, v)) = \begin{vmatrix} \dfrac{dT_1}{dr} & \dfrac{dT_1}{d\theta} \\ \dfrac{dT_2}{dr} & \dfrac{dT_2}{d\theta} \end{vmatrix} = \begin{vmatrix} \cos\theta & -r\sin\theta \\ \sin\theta & r\cos\theta \end{vmatrix} = r(\cos^2\theta + \sin^2\theta) = r.$$

5.7. DOUBLE RIEMANN INTEGRAL

Definition 5.7. The double Riemann integral (Eq. 5.5) of a real-valued function f over a rectangle region $D = [a, b] \times [c, d]$, where $[\overline{x}_i, \overline{y}_j]$ are the small rectangles region D has been divided into and $\overline{x}_i$, $\overline{y}_j$ are the midpoints of the rectangles, is denoted by

$$\iint_D f(x, y) dy\, dx = \lim_{i,j \to \infty} \sum_{i=1}^{n} \sum_{j=1}^{m} f(\overline{x}_i, \overline{y}_j) \Delta A \tag{5.5}$$

The resolution of a double Riemann integral using a $\mathbb{R}^2$ in $\mathbb{R}^2$ mapping affects the integration domain and the function to integrate, simplifying both elements and giving an equivalent solution.

Definition 5.8. For a real-valued function $f: D \subset \mathbb{R}^2 \to M \subset \mathbb{R}$ and a map $T: D^* \subset \mathbb{R}^2 \to D \subset \mathbb{R}^2$, the double Riemann integral Eq. 5.6) is equivalent to

$$\begin{aligned} \iint_D f(x, y)\, dy\, dx &= \int_{x_1}^{x_2} \int_{y_1(x)}^{y_2(x)} f(x, y)\, dy\, dx \\ &= \iint_{D^*} f(x, y) \circ T(u, v)\, |J(T(u, v)|\, dv\, du \\ &= \int_{u_1}^{u_2} \int_{v_1(u)}^{v_2(u)} f(T(u, v))\, |J(T(u, v))|\, dv\, du. \end{aligned} \tag{5.6}$$

Example 5.8. Let the real-valued function $f(x, y) = x^2 y^2$ [43] and the region $D = \{(x, y) \in \mathbb{R}^2 | 1 \le xy \le 2, x \le y \le 4x\}$. (i) Compute a map $T: D^* \subset \mathbb{R}^2 \to D \subset \mathbb{R}^2$. (ii) Compute $J(T(u, v))$. (iii) Draw the regions D and D^*. (iv) Compute the double Riemann integral over the vu-plane.

Solution 5.8. (i) The region D is equivalent to $D = \{(x, y) \in \mathbb{R}^2 \mid 1 \leq xy \leq 2, 1 \leq \frac{y}{x} \leq 4$. If $u = xy$ and $v = \frac{y}{x}$, $x = \sqrt{\frac{u}{v}}$ and $y = \sqrt{uv}$, then $T(u, v) = \left(\sqrt{\frac{u}{v}}, \sqrt{uv}\right), u, v > 0$

(ii) $J(T(u, v)) = \begin{vmatrix} \dfrac{dT_1}{du} & \dfrac{dT_1}{dv} \\[2em] \dfrac{dT_2}{du} & \dfrac{dT_2}{dv} \end{vmatrix} = \begin{vmatrix} \dfrac{\frac{1}{v}}{2\sqrt{\frac{u}{v}}} & \dfrac{-\frac{u}{v^2}}{2\sqrt{\frac{u}{v}}} \\[2em] \dfrac{v}{2\sqrt{uv}} & \dfrac{u}{2\sqrt{uv}} \end{vmatrix} = \dfrac{1}{2v} \neq 0, \forall u, v > 0.$

(iii) See (Fig. **5.5**).

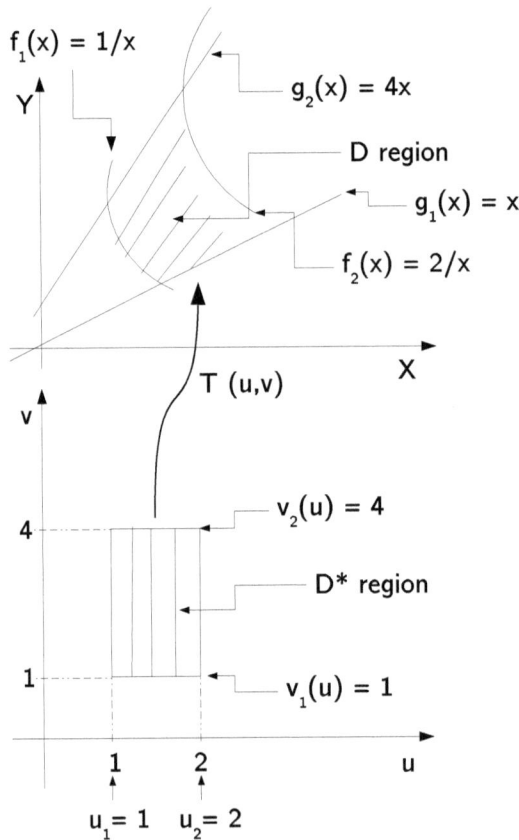

Fig. (5.5). In the xy-plane the independent variable is x and in the vu-plane the independent variable is u.

(iv)$\iint_D f(x,y)\, dy\, dx = \iint_{D^*} f(x,y) \circ T(u,v)\, |J(T(u,v)|\, dv\, du = \int_1^2 \int_1^4 \frac{u^2}{2v}\, dv\, du = \frac{7}{3}\log 2 = 1.6173.$

5.8. ORDER OF INTEGRATION OVER $\mathbb{R}^2$

The transformation of the R-region in the yx-plane into the R^*-region in the xy-plane can be done in two steps (i) rotating the x-axis to point in the negative direction (ii) rotating the x-axis clockwise by 90 degrees (Fig. **5.6**). In this way, the regions R and R^* are equivalent (Eq. 5.7).

$$\begin{aligned}
\iint_R f(x,y)\, dy\, dx &= \int_{x_1}^{x_2} \int_{y_1(x)}^{y_2(x)} f(x,y)\, dy\, dx \\
\iint_{R^*} f(x,y)\, dx\, dy &= \int_{y_1}^{y_2} \int_{x_1(y)}^{x_2(y)} f(x,y)\, dx\, dy
\end{aligned}$$ (5.7)

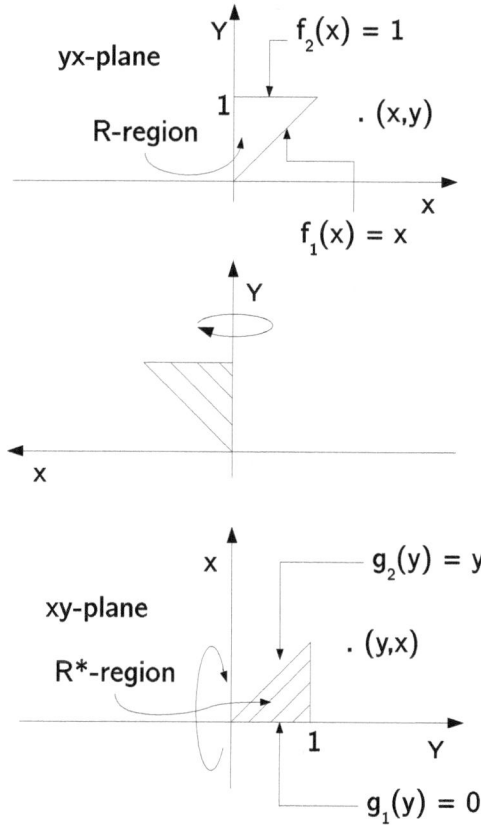

Fig. (5.6). In the yx-plane the independent variable is x and in the xy-plane the independent variable is y.

Example 5.9. Let the real-valued function $f(x,y) = x^2 + y^2$ over the region (Fig. 5.6). (i) Compute the double Riemann integral in the yx-plane. (ii) Compute the double Riemann integral in the yx-plane.

Solution 5.9. (i) $\iint_R f(x,y)\,dy\,dx = \int_0^1 \int_x^1 x^2 + y^2\,dy\,dx = \int_0^1 -\frac{4}{3}x^3 + x^2 + \frac{1}{3}\,dx = \frac{1}{3}$. (ii)$\iint_{R^*} f(x,y)\,dx\,dy = \int_0^1 \int_0^y x^2 + y^2\,dx\,dy = \int_0^1 \frac{1}{3}y^3 + y^3\,dy = \frac{1}{3}$.

The conversion of the planes, implies the inversion of the components of the map, that connects both planes.

Definition 5.9. Let the map $T: D \subset \mathbb{R}^2 \to D \subset \mathbb{R}^2, (T_1(x,y), T_2(x,y))$, where the region D belongs to $y - x$ plane. Then the map $T: D \subset \mathbb{R}^2 \to M \subset \mathbb{R}^2, (T_2(x,y), T_1(x,y))$, where the region M is the transformation of region D in the $x - y$ plane.

Example 5.10. Let the region $D = [0,1] \times [0,2]$ in the $y - x$ plane. What map transforms the region D?

Solution 5.10. $T: [0,1] \times [0,1] \subset \mathbb{R}^2 \to D \subset \mathbb{R}^2, (x, 2y)$, then the map that transforms D in M (in the $x - y$ plane) is $T_{x-y\text{plane}}: [0,1] \times [0,1] \subset \mathbb{R}^2 \to M \subset \mathbb{R}^2, (2y, x)$.

5.9. MAPS $\mathbb{R}^3 \to \mathbb{R}^3$

Definition 5.10. D and D^* are affine spaces in $\mathbb{R}^3$ $T: D^* \to D$ (Fig. **5.7**) of the form $x \mapsto Mx + v,$, where M is a linear transformation on D^* and v is a vector in D [42].

Example   5.11.   For   the   map $T: [0,1] \times [0,2\pi] \times [0,\pi] \subset \mathbb{R}^3 \to \mathbb{R}^3, (\rho\sin\phi\,\cos\theta, \rho\sin\phi\,\sin\theta, \rho\cos\phi)$. (i) What is the domain of T? (ii) What is the image of T?

Solution 5.11. (i) The domain of T in space is the parallelepiped $[0,1] \times [0,2\pi] \times [0,\pi]$. (ii) The image of T is a unit sphere with its centre at the origin of the Cartesian coordinate system.

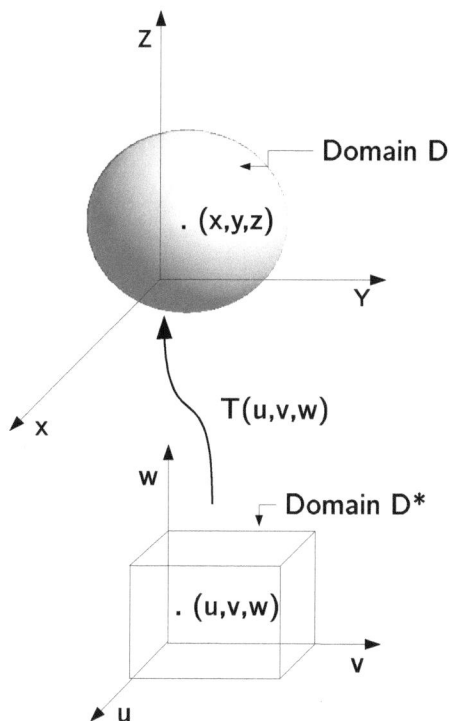

Fig. (5.7). Map $T: D^* \subset \mathbb{R}^3 \to D \subset \mathbb{R}^3$, where $T(D^*) = D$.

5.10. JACOBIAN DETERMINANT ON $\mathbb{R}^3$

Definition 5.11. If T is a transformation such that $T: (u, v, w) \in \mathbb{R}^3 \to (x, y, z) \in \mathbb{R}^3$, the Jacobian determinant (Eq. 5.8) or Jacobian $J(T)$ is the determinant of the **matrix derivatives** of T (Thm. 3.2).

$$J(T(u, v, w)) = \begin{vmatrix} \dfrac{dT_1}{du} & \dfrac{dT_1}{dv} & \dfrac{dT_1}{dw} \\[2mm] \dfrac{dT_2}{du} & \dfrac{dT_2}{dv} & \dfrac{dT_2}{dw} \\[2mm] \dfrac{dT_3}{du} & \dfrac{dT_3}{dv} & \dfrac{dT_3}{dw} \end{vmatrix} \qquad (5.8)$$

Example 5.12. Let T be a transformation such that $T: \mathbb{R}^3 \to \mathbb{R}^3$, $(\rho\sin\phi\cos\theta, \rho\sin\phi$ $\sin\theta, \rho\cos\phi)$. What is the Jacobian determinant?

Solution 5.12.

$$J(T(u,v,w)) = \begin{vmatrix} \dfrac{dT_1}{d\rho} & \dfrac{dT_1}{d\theta} & \dfrac{dT_1}{d\phi} \\[2ex] \dfrac{dT_2}{d\rho} & \dfrac{dT_2}{d\theta} & \dfrac{dT_2}{d\phi} \\[2ex] \dfrac{dT_3}{d\rho} & \dfrac{dT_3}{d\theta} & \dfrac{dT_3}{d\phi} \end{vmatrix} = \begin{vmatrix} \sin\phi\cos\theta & -\rho\sin\phi\sin\theta & \rho\cos\phi\cos\theta \\ \sin\phi\sin\theta & \rho\sin\phi\cos\theta & \rho\cos\phi\sin\theta \\ \cos\phi & 0 & -\rho\sin\phi \end{vmatrix}$$

$$= -\rho^2\sin\phi$$

5.11. TRIPLE RIEMANN INTEGRAL

Definition 5.12. The triple integral (Eq. 5.9) of a real-valued function f over a parallelepiped region $D = [a,b] \times [c,d] \times [e,f]$, where $[\overline{x}_i, \overline{y}_j, \overline{z}_k]$ are the small parallelepipeds region D has been divided into and $\overline{x}_i, \overline{y}_j, \overline{z}_k$ are the midpoints of the parallelepipeds, is denoted by

$$\iiint_D f(x,y,z)dz\,dy\,dx = \lim_{i,j,k\to\infty} \sum_{i=1}^{n}\sum_{j=1}^{m}\sum_{k=1}^{p} f(\overline{x}_i,\overline{y}_j,\overline{z}_k)\Delta V \qquad (5.9)$$

The resolution of a triple Riemann integral using an $\mathbb{R}^3$ in $\mathbb{R}^3$ mapping affects the integration domain and the function to integrate, simplifying both elements and giving an equivalent solution.

Definition 5.13. For a real-valued function $f: D \subset \mathbb{R}^3 \to M \subset \mathbb{R}$ and a map $T: D^* \subset \mathbb{R}^3 \to D \subset \mathbb{R}^3$, the triple Riemann integral is equivalent to (Eq. 5.10)

$$\begin{aligned} \iiint_D f(x,y,z)\,dz\,dy, dx &= \int_{x_1}^{x_2}\int_{y_1(x)}^{y_2(x)}\int_{z_1(x,y)}^{z_2(x,y)} f(x,y,z)\,dz\,dy\,dx \\ &= \iiint_{D^*} f(x,y,z)\circ T(u,v,w)\,|J(T(u,v,w)|\,dw\,dv\,du \qquad (5.10) \\ &= \int_{u_1}^{u_2}\int_{v_1(u)}^{v_2(u)}\int_{w_1(u,v)}^{w_2(u,v)} f(T(u,v,w))\,|J(T(u,v,w))|\,dw\,dv\,du. \end{aligned}$$

Example 5.13. Let the real-valued function $f(x,y,z) = \sqrt{x^2 + y^2 + z^2}$ and the region $D = \{(x,y,z) \in \mathbb{R}^3 \,|\, x^2 + y^2 + z^2 \leq 1 \text{ such that } x,y,z \in \mathbb{R}\}$. (i) Compute map $T: D \subset \mathbb{R}^3 \to D^* \subset \mathbb{R}^3$. (ii) Compute $J(T(\rho,\theta,\phi))$. (iii) Draw the regions D and D^*. (iv) Compute the triple Riemann integral over a region in space.

Solution 5.13. (i) Region D is a solid unit sphere. So $T: [0,1] \times [0,2\pi] \times [0,\pi] \subset \mathbb{R}^3 \to \mathbb{R}^3, (\rho\sin\phi\cos\theta, \rho\sin\phi \sin\theta, \rho\cos\phi)$.

(ii)

$$J(T(u,v,w)) = \begin{vmatrix} \dfrac{dT_1}{d\rho} & \dfrac{dT_1}{d\theta} & \dfrac{dT_1}{d\phi} \\[2mm] \dfrac{dT_2}{d\rho} & \dfrac{dT_2}{d\theta} & \dfrac{dT_2}{d\phi} \\[2mm] \dfrac{dT_3}{d\rho} & \dfrac{dT_3}{d\theta} & \dfrac{dT_3}{d\phi} \end{vmatrix}$$

$$= \begin{vmatrix} \sin\phi\cos\theta & -\rho\sin\phi\sin\theta & \rho\cos\phi\cos\theta \\ \sin\phi\sin\theta & \rho\sin\phi\cos\theta & \rho\cos\phi\sin\theta \\ \cos\phi & 0 & -\rho\sin\phi \end{vmatrix}$$

$$= -\rho^2\sin\phi$$

(iii) See (Fig. **5.8**)

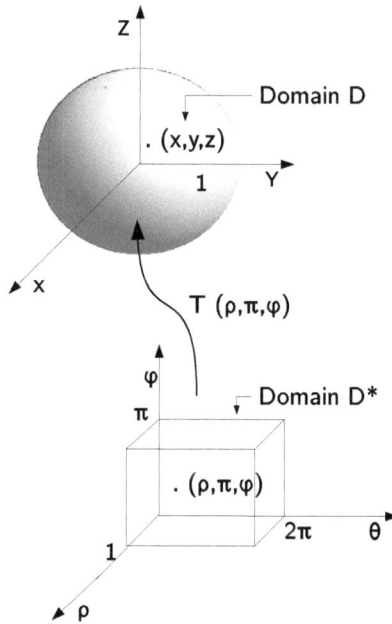

Fig. (5.8). D Region domain of the function $f(x,y) = \sqrt{x^2 + y^2 + z^2}$.

(iv) $\iiint_D f(x,y,z) \, dz \, dy \, dx = \iiint_{D^*} f(x,y,z) \circ T(\rho,\theta,\phi) \, |J(T(\rho,\theta,\phi)| \, d\rho \, d\theta \, d\phi$

$$= \int_0^\pi \int_0^{2\pi} \int_0^1 \rho^2 \, \rho^2 \sin\phi d\rho \, d\theta, d\phi = 2\pi^2 \int_0^1 \rho^4 \sin\phi d\rho = \frac{4}{5}\pi.$$

5.12. ORDER OF INTEGRATION OVER $\mathbb{R}^3$

There are six transformations of the R-region in the $\mathbb{R}^3$ space (Eq. 5.11) and each one corresponds to a different Cartesian coordinate system (Fig. **5.9**).

$$\int_{x_1}^{x_2} \int_{y_1(x)}^{y_2(x)} \int_{z_1(x,y)}^{z_2(x,y)} f(x,y,z) \, dz \, dy \, dx =$$

$$= \int_{y_1}^{y_2} \int_{x_1(y)}^{x_2(y)} \int_{z_1(y,x)}^{z_2(y,x)} f(x,y,z) \, dz \, dx \, dy$$

$$= \int_{x_1}^{x_2} \int_{z_1(x)}^{z_2(x)} \int_{y_1(z,x)}^{y_2(z,x)} f(x,y,z) \, dy \, dz \, dx$$

$$= \int_{z_1}^{z_2} \int_{x_1(z)}^{x_2(z)} \int_{y_1(x,z)}^{y_2(x,z)} f(x,y,z) \, dy \, dx \, dz \qquad \textbf{(5.11)}$$

$$= \int_{z_1}^{z_2} \int_{y_1(z)}^{y_2(z)} \int_{x_1(y,z)}^{x_2(y,z)} f(x,y,z) \, dx \, dy \, dz$$

$$= \int_{y_1}^{y_2} \int_{z_1(y)}^{z_2(y)} \int_{x_1(z,y)}^{x_2(z,y)} f(x,y,z) \, dx \, dz \, dy.$$

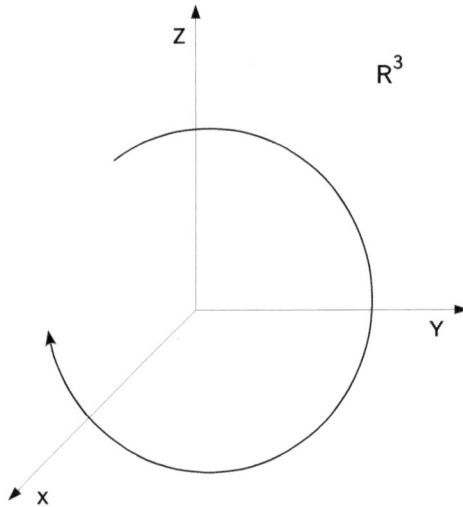

Cartesian coordinate system oriented in the clockwise direction

Fig. (5.9). The triple integral whose order of differentials is $dz, dy \, dx$ corresponds to the order of the axes $Y \rightarrow X \rightarrow Z$. This order is obtained when reading the Cartesian coordinate system in the clockwise direction.

Example 5.14. Let the region D be a parallelepiped whose edges in the ZYX-space are $4 \times 5 \times 6$. Draw its six representations in three-dimensional space, in order of the differentials described in the six triple Riemann integrals (Eq. 5.11).

Solution 5.14. (i) $\iiint_D f(x, y, z) dz dy dx$ corresponds to $Z \to Y \to X$ (Fig. **5.14**(a)).

(ii) $\iint_D f(x, y, z) dz dx dy$ corresponds to $Z \to X \to Y$ (Fig. **5.14**(b)).

(iii) $\iiint_D f(x, y, z) dy dz dx$ corresponds to $Y \to Z \to X$ (Fig. **5.14**(c)).

(iv) $\iiint_D f(x, y, z) dy\, dx dz$ corresponds to $Y \to X \to Z$ (Fig. **5.14**(d)).

(v) $\iiint_D f(x, y, z) dx dy dz$ corresponds to $Y \to X \to Z$ (Fig. **5.14**(e)).

(vi) $\iint_D f(x, y, z) dx dz dy$ corresponds to $X \to Z \to Y$ (Fig. **5.14**(f)).

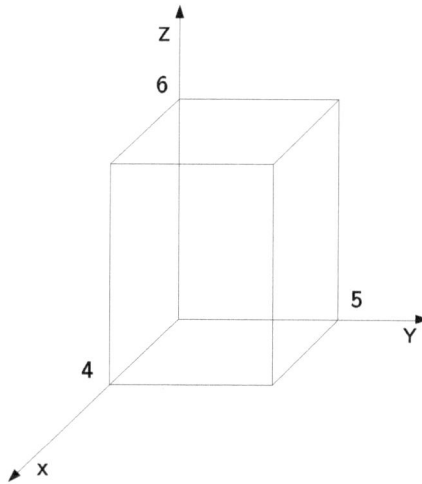

$(a) ZYX$-space in the clockwise orientation.

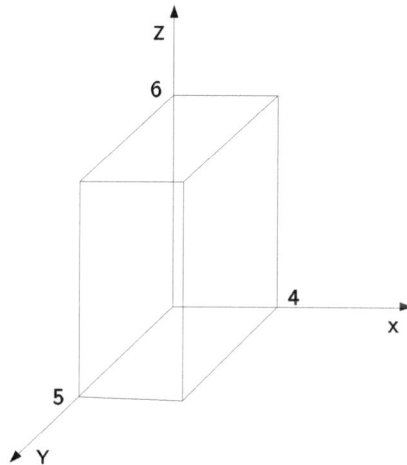

(b) ZXY-space in the clockwise orientation.

Fig. (5.10). Representations in $\mathbb{R}^3$ of the ZYX and ZXY cases.

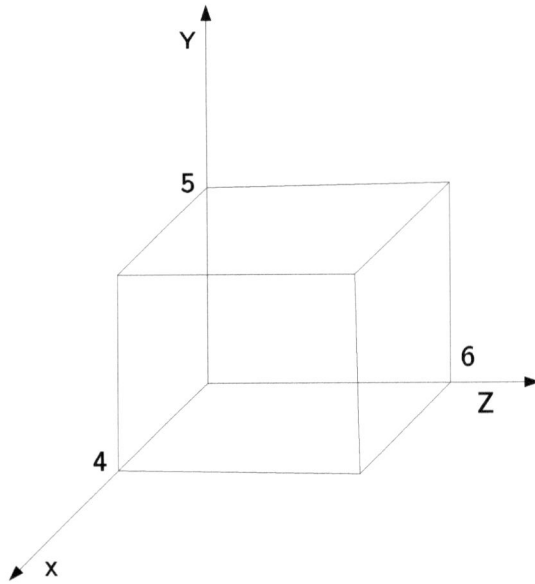

(c) YZX-space in the clockwise orientation.

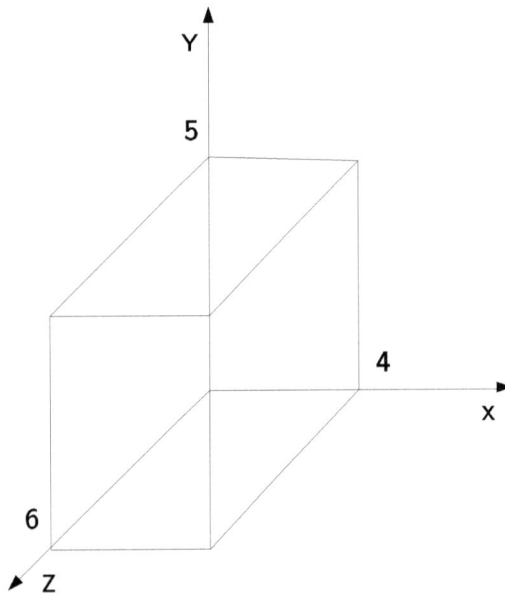

(d) YXZ-space in the clockwise orientation.

Fig. (5.11). Representations in $\mathbb{R}^3$. YZX and YXZ cases.

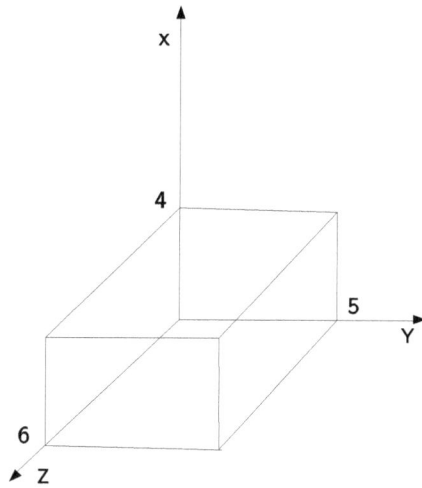

(e) XYZ-space in the clockwise orientation.

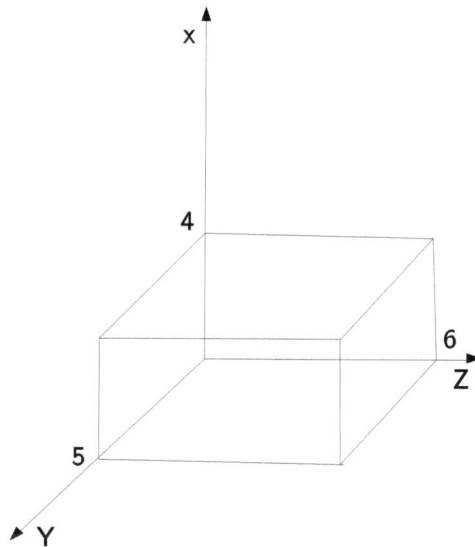

(f) XZY-space in the clockwise orientation.

Fig. (5.12). Representations in $\mathbb{R}^3$ of the XYZ and XZY cases.

5.13. CASE STUDY: PROBLEM ON A DOUBLE INTEGRAL OVER A CIRCLE

Case 5.1. Let D be a circle with the radius centred at the origin.

Evaluate

$$\iint_D (1 - x^2 - y^2)dxdy.$$

Case adapted with permission of the author [9].

For convenience, we will let $I = \iint_D (1 - x^2 - y^2)dxdy$. If a problem involves circular geometry or terms like $x^2 + y^2$), it might be convenient to use polar coordinates. We note that the integrand $1 - x^2 - y^2$ can be written $1 - (x^2 + y^2)$. Hence, we identify the pattern and change to polar coordinates. Recall:

In polar coordinates, $x = r\cos\theta$ and $y = r\sin\theta$. Thus, $x^2 + y^2 = r^2$. The differential area is element $dA = rdrd\theta$.

We can now write the integrand as

$$1 - x^2 - y^2 = 1 - (x^2 + y^2) = 1 - r^2.$$

Rewriting the double integral in polar coordinates, we get

$$I = \iint_D (1 - x^2 - y^2)dxdy = \int_{r=0}^{a} \int_{\theta=0}^{2\pi} (1 - r^2)rd\theta dr \qquad (5.12)$$

Recall:

If the bounds of a double integral are constants and integrand factors, then the integral factors are

$$\int_{r=a}^{b} \int_{\theta=c}^{d} f(r)g(\theta)d\theta dr = \int_{r=a}^{b} f(r)dr \int_{\theta=c}^{d} g(\theta)d\theta \qquad (5.13)$$

Hence, it can be expressed

$$I = \int_0^a (1 - r^2)rdr \int_0^{2\pi} d\theta \qquad (5.14)$$

$$= 2\pi \int_0^a (r - r^3)dr \qquad (5.15)$$

$$= 2\pi(\frac{1}{2}a^2 - \frac{1}{4}a^4) \qquad (5.16)$$

After simplifying, we get

$$\iint_D (1 - x^2 - y^2)dxdy = \frac{1}{2}\pi(2a^2 - a^4).$$

5.14. CASE STUDY: FUNCTION WITH NO ELEMENTARY ANTIDERIVATIVE

Case 5.2. Evaluate the double integral

$$\int_0^2 \int_{\frac{y}{2}}^1 ye^{x^3}\, dx\, dy.$$

Case adapted with permission of the author [10]

Since function $f(x) = e^{x^3}$ does not have an antiderivative, the iterated integral cannot be calculated as it is. Therefore, we need to reverse the order of integration and the region of integration $\{(x,y) \in \mathbb{R}^2 | 0 \le y \le 2, \frac{y}{2} \le x \le 1\}$ of both inequalities. Assuming that $0 < x < 1$ (Note 14) we obtain from the first inequality $0 \le y$ and from the second $y \le 2x$, thus $0 \le y \le 2x$. Now reversing the order of integration we have

$$\int_0^2 \int_{\frac{y}{2}}^1 ye^{x^3}\, dx\, dy$$

$$= \int_0^1 \int_0^{2x} ye^{x^3}\, dy\, dx$$
$$= \int_0^1 \left(\frac{y^2}{2}e^{x^3}\right)\Big|_0^{2x}$$
$$= 2\int_0^1 x^2 e^{x^3}\, dx$$
$$= \frac{2}{3}\int_0^1 e^u\, du \text{ (Note 5.2)}$$
$$= \frac{2}{3}(e - 1).$$

Note 5.1. To determine the maximal domain of x we see what value of y maximises the range of the variable x. In this case, for $y = 0$ we have $0 \le x \le 1$, which is the largest range for x; and for $y = \frac{3}{2}$ we have $\frac{3}{4} \le x \le 1$, which is a subset of $0 \le x \le 1$.

Note 5.2. If $u = x^3 \Rightarrow du = 3x^2\, dx \Rightarrow dx = \frac{du}{3x^2}$.

5.15. EXERCISES

Exercise 5.1. Let the real-valued function $f(x) = x^2\sqrt{x^3}$ and the interval $[1,2] \in \mathbb{R}$. (i) What is the area under the curve in this domain interval? (ii) What map is convenient to simplify this integral? (iii) Solve integral (i) using the map $T(u)$. (iv) Solve the integral using the change of variable method $u = x^3$. (v) Analyse it geometrically.

Exercise 5.2. Let the real-valued function $f(x,y) = x^2 + y^2$ and the domain region the unit circle. (i) What is the volume under the graph on that domain region? (ii) What map is convenient to simplify this integral? (iii) Solve the integral using the map. (iv) Analyse the domain region geometrically.

Exercise 5.3. Let the real-valued function $f(x,y) = 2 - 3x + xy$ and the domain region the triangle spanned by $(0,0)$, $(1,0)$, $(1,3)$ [44]. (i) What is the volume under the graph of the domain region in the xy-plane? (ii) What is the volume under the graph of the domain region in the yx-plane? (iii) Analyse the domain region geometrically.

Exercise 5.4. Let the real-valued functions $f(x,y) = 4 - x^2 - y^2$ and $f(x,y) = \sqrt{x^2 + y^2}$. (i) Determine the volume between both functions. (ii) What map is convenient to simplify this integral? (iii) Solve the integral using map (ii).

Exercise 5.5. Consider a cylinder with radius $r = 1$ and height $z = 2$. (i) Determine the volume of the cylinder by planar geometry. (ii) What map is convenient to simplify the triple integral? (iii) Use a triple integral to determine its volume. (iv) Analyse the domain region and the figure geometrically.

CHAPTER 6

Integration Over Unbounded Regions

Abstract: This section reviews the integrals whose domain of integration corresponds to an unbounded region, or a partially unbounded region, on a plane or in space. This type of integrals are called improper integrals.

Keywords: Bounded and not closed regions, Closed and not bounded regions, Convergent integral, Divergent integral, Improper integral on a plane, Improper Integral in space, Indeterminations over a region, Indetermination over a space, Lower bounded region, Non-continuous function, Not closed and Not bounded regions, Riemman integral, Unbounded regions, Unbounded function, Upper bounded regions.

6.1. UNBOUNDED REGIONS

Definition 6.1. A bounded region is a region that has an upper and a lower bound. An unbounded set would have the opposite characteristics, its upper and/or lower bounds would not be finite [45].

Example 6.1. Provide examples of closed and bounded sets.

Solution 6.1. Closed and bounded: [1,2]. Closed and not bounded: $\cup_{n \in z} [4n, 4n + 1]$. Bounded and not closed: (1,2). Not closed and not bounded: $\cup_{n \in z} (4n^2, 4n^2 + 1)$ [46].

A region in space can be transformed into another similar region by translating its perimeter (Fig. **6.1**).

6.1.1. Improper Integrals on a Plane

Definition 6.2. A double integral is an **improper integral** if the region of integration is not finite, *i.e.* if the integration approaches ∞ or $-\infty$, or if the function to integrate is non-continuous (unbounded) on the integration interval [47].

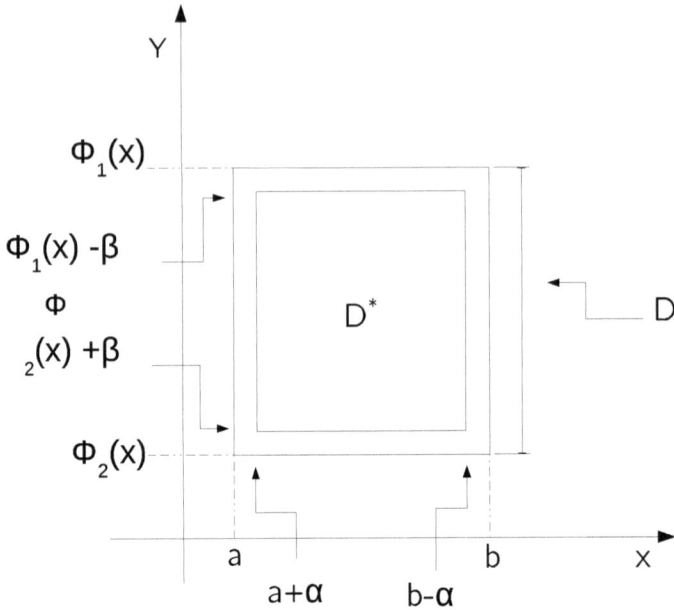

Fig. (6.1). If $\alpha, \beta \to 0 \Rightarrow D^* \to D$. Contrarily, if $\alpha, \beta \to \varepsilon > 0 \Rightarrow D \to D^*$.

Note 6.1. A function that does not have a maximum or minimum x-value is called unbounded.

If f is continuous and bounded on the $D_{\alpha,\beta}$ region, the double Riemann integral (Eq. 6.1) exists.

$$\lim_{(\alpha,\beta)\to(0,0)} \iint_{D_{\alpha,\beta}} f \, dA. \tag{6.1}$$

Example 6.2. Let the real-valued function $f(x,y) = \frac{1}{\sqrt{xy}}$, where the D-region is $[0,1] \times [0,1]$. (i) Are there indeterminations in the domain region? (ii) Evaluate the double integral as an improper integral.

Solution 6.2. (i) Yes, the indeterminations are $(x,0)$ and $(0,y)$ in the xy −plane. Then, the dynamics will be to construct a succession of squares with edge at point $(1,1)$, whose areas grow in the direction of the arrow (Fig. **6.2**).

(ii) $\int_0^1 \int_0^1 \frac{1}{\sqrt{xy}} \, dy dx = \lim(\alpha,\beta) \to (0,0) \int_\beta^1 \int_\alpha^1 \frac{1}{\sqrt{xy}} \, dy dx =$
$\lim_{\alpha \to 0} \int_\alpha^1 \frac{1}{\sqrt{y}} \, dy \lim_{\beta \to 0} \int_\beta^1 \frac{1}{\sqrt{x}} \, dx = \lim_{(\alpha,\beta)\to(0,0)} 2 - 2\sqrt{\beta} + 2 - 2\sqrt{\alpha} = 4.$

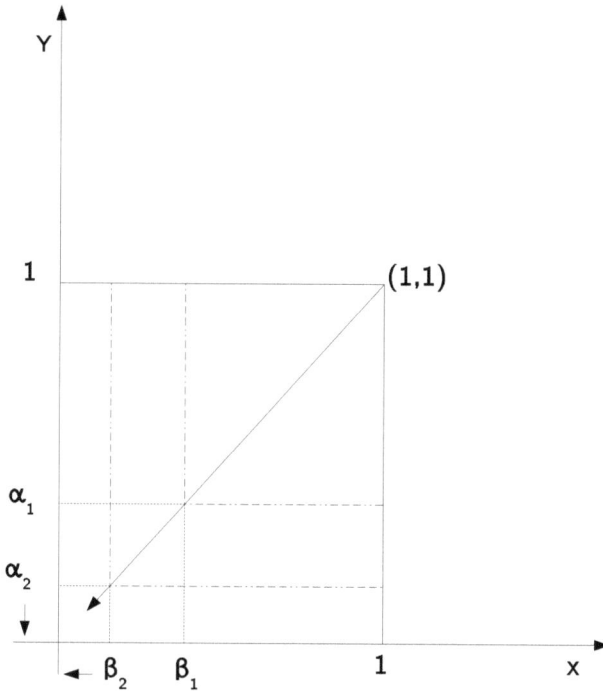

Fig. (6.2). The indeterminations over the D region are the lines $(x, 0)$ and $(0, y)$.

6.1.2. Improper Integrals in Space

Definition 6.3. A triple integral is an **improper integral** if the region of integration is not finite, *i.e.* if the integration approaches ∞ or $-\infty$, or if the function to integrate is non-continuous (unbounded) on the integration interval [47].

Note 6.2. A function that does not have a maximum or minimum x-value is called unbounded.

If f is continuous and bounded on the $D_{\alpha,\beta,\gamma}$ region, the triple Riemann integral (Eq. 6.2) exists.

$$\lim_{(\alpha,\beta,\gamma)\to(0,0,0)} \iiint_{D_{\alpha,\beta,\gamma}} f \, dV. \tag{6.2}$$

Example 6.3. Let the real-valued function $f(x, y) = \frac{1}{\sqrt{xyz}}$, where the D-region is $[0,1] \times [0,1] \times [0,1]$. (i) Are there indeterminations in the domain region? (ii) Evaluate the triple integral as an improper integral.

Solution 6.3. (i) Yes, the indeterminations are $(x,0,0),(0,y,0)$ and $(0,0,z)$ in the xyz −space. Then, the dynamics will be to construct a succession of cubes with edge at point $(1,1,1)$, whose volumes grow in the direction of the arrow. (Fig. **6.3**).
(ii)

$$\int_0^1\int_0^1\int_0^1 \frac{1}{\sqrt{xyz}}\,dzdydx = \lim(\alpha,\beta,\gamma)\to(0,0,0)\int_\beta^1\int_\alpha^1\int_\gamma^1 \frac{1}{\sqrt{xyz}}\,dzdy\,dx$$

$$= \lim_{\beta\to0}\int_\beta^1 \frac{1}{\sqrt{z}}\,dz[\lim_{\alpha\to0}\int_\alpha^1 \frac{1}{\sqrt{y}}\,dy][\lim_{\gamma\to0}\int_\gamma^1 \frac{1}{\sqrt{x}}\,dx]$$

$$= \lim_{(\alpha,\beta,\gamma)\to(0,0,0)}6 - 2(\sqrt{\beta}+\sqrt{\alpha}+\sqrt{\gamma}) = 8.$$

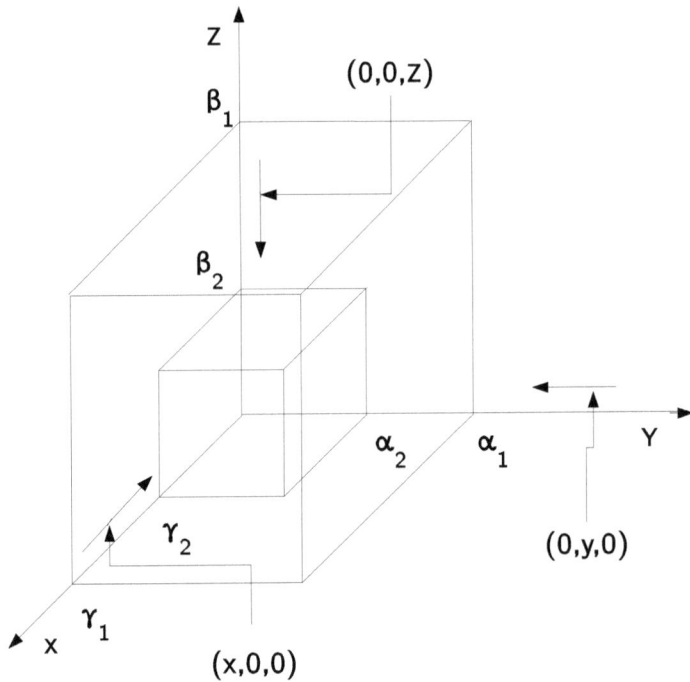

Fig. (6.3). The indeterminations over the D region are the lines $(x,0,0)$, $(0,y,0)$ and $(0,0,z)$.

6.2. CASE STUDY: IMPROPER INTEGRAL – I

Compute the improper integral $\int_1^3 \frac{3}{x^2-3x}\,dx$. **Case adapted with permission of the author** [11].

This function has a vertical asymptote at $x = 3$, where the $x^2 - 3x$ becomes undefined but is otherwise continuous on the interval $[1, 3]$. We will eventually apply (Def. 6.2). It makes sense to do the integration of partial fractions without limits. The degree of the numerator is less than the degree of the denominator $0 < 2$ and the denominator factors into distinct linear factors are: $x(x - 3)$.

$$\int_1^3 \frac{3}{x^2 - 3x} = \frac{3}{x(x-3)} = \frac{A}{x} + \frac{B}{x-3} = \frac{Ax - 3A + Bx}{x^2 - 3x}$$

Comparing the numerators of the first and last functions and solving for A and B gives $0 = A + B$ and $3 = -3A \Rightarrow A = -1, B = 1$. Hence,

$$\int_1^3 \frac{3}{x^2 - 3x}\, dx = \int_1^3 \left(\frac{1}{x-3} - \frac{1}{x}\right)\, dx = \ln|x - 3| - \ln|x| + c$$
$$= \ln\left|\frac{x - 3}{x}\right| + c.$$

Notice how it has been simplified. So applying (Def. 6.2)

$$\int_1^3 \frac{3}{x^2 - 3x}\, dx$$

$$= \lim_{c \to 3^-} \int_1^c \frac{3}{x^2 - 3x}\, dx$$

$$= \lim_{c \to 3^+} \left[\ln\left|\frac{x - 3}{x}\right|\right]_1^c$$

$$= \lim_{c \to 3^+} \left[\ln\left|\frac{c-3}{c}\right| - \ln\left|\frac{1-3}{1}\right|\right] \tag{6.3}$$

$$= \infty.$$

The integral diverges.

6.3. CASE STUDY: IMPROPER INTEGRAL – II

Compute the improper integral $\int_0^\infty \frac{1}{(x-5)^{\frac{1}{3}}}\, dx$. **Case adapted with permission of the author** [12].

This integral is improper for two reasons: A) the integrand $f(x) = \dfrac{1}{(x-5)^{\frac{1}{3}}}$ is undefined at $x = 5$, which is on the interval of integration; B) one of the limits of integration is infinite. First, we need to break the integral into two separate integrals at $x = 5$:

$$\int_0^\infty \frac{1}{(x-5)^{\frac{1}{3}}}\,dx = \int_0^5 \frac{1}{(x-5)^{\frac{1}{3}}}\,dx + \int_5^\infty \frac{1}{(x-5)^{\frac{1}{3}}}\,dx.$$

In the second integral, the lower limit $x = 5$ makes the integrand undefined and the upper limit is infinite. Thus, we need to break the second integral into two parts. We can choose any point between 5 and ∞ as the breaking point; we will use $x = 6$.

$$\int_0^5 \frac{1}{(x-5)^{\frac{1}{3}}}\,dx + \int_5^\infty \frac{1}{(x-5)^{\frac{1}{3}}}\,dx$$

$$= \int_0^5 \frac{1}{(x-5)^{\frac{1}{3}}}\,dx + \int_5^6 \frac{1}{(x-5)^{\frac{1}{3}}}\,dx +$$

$$\int_6^\infty \frac{1}{(x-5)^{\frac{1}{3}}}\,dx. \tag{6.4}$$

Then, we will compute the three integrals. Note that $\int \dfrac{1}{(x-5)^{\frac{1}{3}}}\,dx = \dfrac{3}{2}(x-5)^{\frac{2}{3}} + C$. First,

$$\int_0^5 \frac{1}{(x-5)^{\frac{1}{3}}}\,dx + \int_5^\infty \frac{1}{(x-5)^{\frac{1}{3}}}\,dx$$

$$= \lim_{a \to 5^-} \int_0^a \frac{1}{(x-5)^{\frac{1}{3}}}\,dx$$

$$= \lim_{a \to 5^-} \left[\frac{3}{2}(x-5)^{\frac{2}{3}} \right]_0^a$$

$$= \lim_{a \to 5^-} \left[\frac{3}{2}(a-5)^{\frac{2}{3}} - \frac{3}{2}(-5)^{\frac{2}{3}} \right]$$

$$= 0 - \frac{3}{2}(-5)^{\frac{2}{3}}$$

$$= -\frac{3}{2}(5)^{\frac{2}{3}}. \tag{6.5}$$

Note 6.3 $(-5)^{2/3} = 5^{2/3}$ the even power "2" eliminates the minus sign (Eq. 6.5).

$$\int_5^6 \frac{1}{(x-5)^{\frac{1}{3}}} \, dx = \lim_{b \to 5^+} \int_b^6 \frac{1}{(x-5)^{\frac{1}{3}}} \, dx$$

$$= \lim_{b \to 5^+} \left[\frac{3}{2}(x-5)^{\frac{2}{3}} \right]_b^6$$

$$= \lim_{b \to 5^+} \left(\frac{3}{2} - \frac{3}{2}(b-5)^{\frac{2}{3}} \right)$$

$$= \frac{3}{2} - 0$$

$$= \frac{3}{2}. \tag{6.6}$$

Finally,

$$\int_6^\infty \frac{1}{(x-5)^{\frac{1}{3}}} \, dx = \lim_{c \to \infty} \int_6^c \frac{1}{(x-5)^{\frac{1}{3}}} \, dx$$

$$= \lim_{c \to \infty} \left[\frac{3}{2}(x-5)^{\frac{2}{3}} \right]_6^c$$

$$= \lim_{c \to \infty} \left(\frac{3}{2}(c-5)^{\frac{2}{3}} - \frac{3}{2} \right)$$

$$= \infty - \frac{3}{2}$$

$$= \infty. \tag{6.7}$$

The first two integrals converged, but the third diverged to ∞ . Therefore,

$$\int_0^\infty \frac{1}{(x-5)^{\frac{1}{3}}}\, dx. = \infty.$$

6.4. EXERCISES

Exercise 6.1. Let the real-valued function $f(x,y) = \dfrac{1}{xy}$ and the region D be defined by $0 \leqslant x \leqslant 1$, and $0 \leqslant y \leqslant 1$. (i) Solve the integral. (ii) Discuss the implications of dealing with an improper integral. (iii) Draw the domain region.

Exercise 6.2. Let the real-valued function $f(x,y) = \dfrac{1}{xy}$ and the region D be defined by $1 \leqslant x \leqslant \infty$, and $1 \leqslant y \leqslant \infty$. (i) Solve the integral. (ii) Discuss the implications of dealing with an improper integral. (iii) Draw the domain region.

Exercise 6.3. Let the real-valued function $f(x,y) = \dfrac{1}{x-y}$ and the region D be defined by $0 \leqslant x \leqslant 1$, and $0 \leqslant y \leqslant 1$. (i) Solve the integral. (ii) Discuss the implications of dealing with an improper integral. (iii) Draw the domain region.

Exercise 6.4. Let the real-valued function $f(x,y) = \dfrac{1}{xy}$ and the region D be defined by $1 \leqslant x \leqslant 2$, and $1 \leqslant y \leqslant 2$. (i) Solve the integral as an improper integral. (ii) Discuss the implications of dealing with an improper integral. (iii) Draw the domain region.

Exercise 6.5. Discuss the convergence of the following improper integrals [48]:

$$\lim_{\alpha \to \infty} \int_2^\alpha \frac{1}{x^3}\, dx = \frac{1}{8} \text{ and } \lim_{\alpha \to \infty} \int_\alpha^{-1} \frac{1}{x}\, dx = -\infty.$$

CHAPTER 7

Integration Over Curves and Surfaces

Abstract: This chapter introduces the Line and Surface Integrals as a measure of the effect of a vector field or scalar field over oriented curves and surfaces.

Keywords: Curves, line integral of scalar functions, line integral of vector functions, maps, mass of a wire, oriented line, oriented region, oriented surface, oriented closed surface, oriented trajectory, scalar field, surfaces, surface integral of scalar functions, surface integral of vector functions, vector field.

7.1. CURVES AND SURFACES

Not all curves or surfaces are graphs of a function (Fig. **7.1**). A function is restricted to an element of its domain that corresponds to a single element of its image. A map is not limited by this constraint.

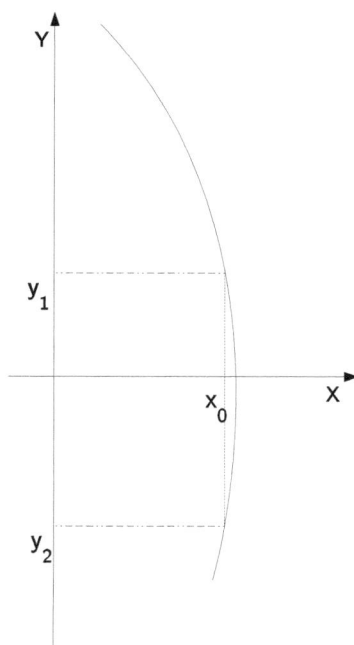

Fig. (7.1). This graph is not a graph of a function because x_0 has two elements y_1 and y_2 in the image.

7.2. CURVES

7.2.1. Maps $\mathbb{R} \rightarrow \mathbb{R}^2$

A curve, as the unit circle, does not come from a function for this reason: its graph is divided into two regions, its upper curve ($f_1(x) = \sqrt{1 - x^2}$) and its lower curve ($f_2(x) = -\sqrt{1 - x^2}$). However, we can use a map from $\mathbb{R}$ to $\mathbb{R}^2$ to get the same figure $T(\theta) = (\cos\theta, \sin\theta), \theta \in [0, 2\pi]$ (Fig. **7.2**). The construction of a map depends on the graph you want to get. If this graph comes from a function then the map will have the form $T(t) = (t, f(t)), t \in [t_0, t_1]$. If the graph does not come from a function, then it would be necessary to build the map from scratch, where $T(t) = (T_1(t), T_2(t)), t \in [t_0, t_1]$.

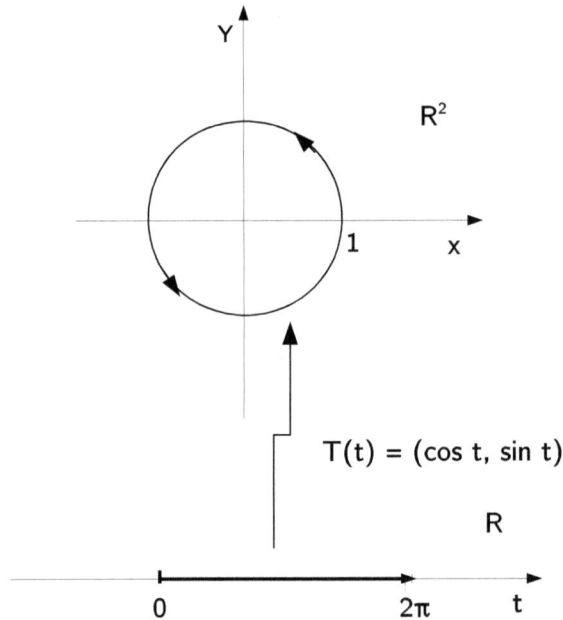

Fig. (7.2). The map transforms the oriented segment into the oriented perimeter of the unit circle.

Note 7.1. Two maps are equivalent if both represent the same trajectory and their norms (Def. 1.4) are equal.

7.2.2. Maps $\mathbb{R} \rightarrow \mathbb{R}^3$

An oriented curve in a space is built using a map $T(t) = (T_1(t), T_2(t), T_3(t)), t \in [t_0, t_1]$ (Fig. **7.3**). In this sense, this map is an extension of a map from $\mathbb{R}$ to $\mathbb{R}^3$. As you can see, it is not possible to build the map from a function whose graph is that curve.

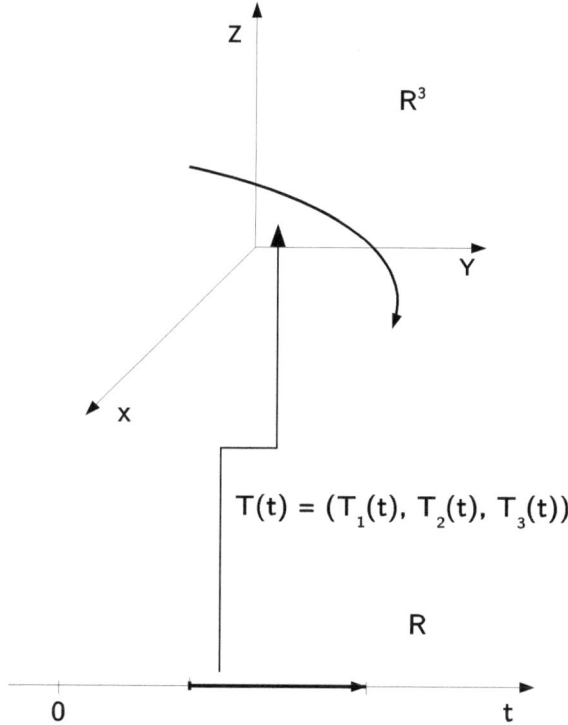

Fig. (7.3). The oriented curve in a space corresponds to the oriented segment in a space $\mathbb{R}$.

Note 7.2. A curve can be oriented indicating with an arrow its orientation or using the notation $t \in [t_1, t_2]$. In this case, t_1 is the starting point of the segment and t_2 is the final point of the segment.

Note 7.3. Two maps are equivalent if both maps represent the same trajectory and their norms (Def. 1.4) are equal.

7.2.3. Line Integral of a Scalar Function

Definition 7.1. A Line Integral of a Scalar Function (Eq. 7.1) measures the total value of a scalar magnitude (Sect. 7.2.3) represented by the real-valued function $f: \mathbb{R}^n \to \mathbb{R}$, over the **non-oriented** line represented by the map $c: \mathbb{R} \to \mathbb{R}^n$.

$$\int_C f \circ T(t) \|T'(t)\| dt = \int_a^b f(T(t)) \|T'(t)\| dt, \text{where } t \in [a, b] \subset \mathbb{R}. \qquad \textbf{(7.1)}$$

Note 7.4. The orientation of the curve **is not** taken into account in this integral.

Example 7.1. Consider a map of a wire $T(t) = (\cos t, \sin t), t \in [0, \pi]$ and be the density of the wire given by the real-valued function $f(x, y) = x^3 + y^3$. Compute the mass of the wire.

Solution 7.1. $\int_T f \circ T(t) \, ||T'(t)|| \, dt = \int_0^\pi (\cos^3 t + \sin^3 t)\sqrt{\cos^2 t + \sin^2 t} \, dt = \frac{4}{3}$. If f were given in *grams/cm* and $T(t)$ were given in *cm*, then the mass of the wire would be $\frac{4}{3}$ *grams*.

Example 7.2. As a particular case, if the real-valued function f is equal to 1, then the Line Integral of a Scalar Function represents the **length** of curve T. Let the trajectory $T(t) = (\cos t, \sin t), t \in [0, 2\pi]$ and the real-valued function $f(x, y) = 1$. Compute the line integral of scalar function.

Solution 7.2. $\int_T ||T'(t)|| \, dt = \int_0^{2\pi} \sqrt{-\sin^2 t + \cos^2 t} \, dt = 2\pi.$

7.2.4. Line Integral of a Vector Function

Definition 7.2. A Line Integral of a Vector Function measures the effect of a vector field $F: \mathbb{R}^n \to \mathbb{R}^n$ (Sect. 4.1) on an **oriented** trajectory $T: \mathbb{R} \to \mathbb{R}^n$ (Eq. 7.2). From the physical point of view, the line integral of a vector function measures the **work** done to move a particle on an oriented trajectory T with the influence of a vector field F.

$$\oint_T F \circ T(t) \cdot T'(t) \, dt = \int_{t_1}^{t_2} F(T(t)) \cdot T'(t) \, dt, \text{where} t \in [t_1, t_2] \subset \mathbb{R}. \qquad (7.2)$$

Example 7.3. A particle moves along the oriented trajectory $T(t) = (\cos t, \sin t), t \in [0, 2\pi]$ and the force is represented by the vector field $F(x, y) = (-y, x)$ (Fig. **7.2**). (i) Compute the work done by the force field on a particle that moves along curve T. (ii) Compute the same line integral of vector function in the opposite trajectory.

Solution 7.3. (i) $\oint_T F \circ T(t) \cdot T'(t) \, dt = \int_0^{2\pi} (-\sin t, \cos t) \cdot (-\sin t, \cos t) \, dt = \int_0^{2\pi} dt = 2\pi$. The work in this direction is positive. (ii) $\oint_T F \circ T(t) \cdot T'(t) \, dt = \int_{2\pi}^0 (-\sin t, \cos t) \cdot (-\sin t, \cos t) \, dt = \int_{2\pi}^0 dt = -2\pi$. The work in this direction is negative.

7.3. SURFACES

7.3.1. Maps $\mathbb{R}^2 \to \mathbb{R}^3$

As an example, a surface representing a unit sphere circle does not come from a function for this reason: its graph is divided into two regions, its upper outer surface $f_1(x,y) = \sqrt{1 - x^2 - y^2}$ and its lower outer surface $f_2(x,y) = -\sqrt{1 - x^2 - y^2}$. However, we can use a map from $\mathbb{R}^2$ to $\mathbb{R}^3$ to get the same figure $T(\theta,\phi) = (\sin\phi\cos\theta, \sin\phi\sin\theta, \cos\phi), \theta \in [0,2\pi], \phi \in [0,\pi]$ (Fig. **7.4**). The construction of a map depends on the graph you want to get. If this graph comes from a function then the map will have the form $T(s,t) = (s,t,f(s,t)), (s,t) \in [s_0,s_1] \times [t_0,t_1]$. If the graph does not come from a function, then it will be necessary to build it from scratch, where $T(s,t) = (T_1(s,t), T_2(s,t), T_3(s,t)) \ (s,t) \in [s_0,s_1] \times [t_0,t_1]$.

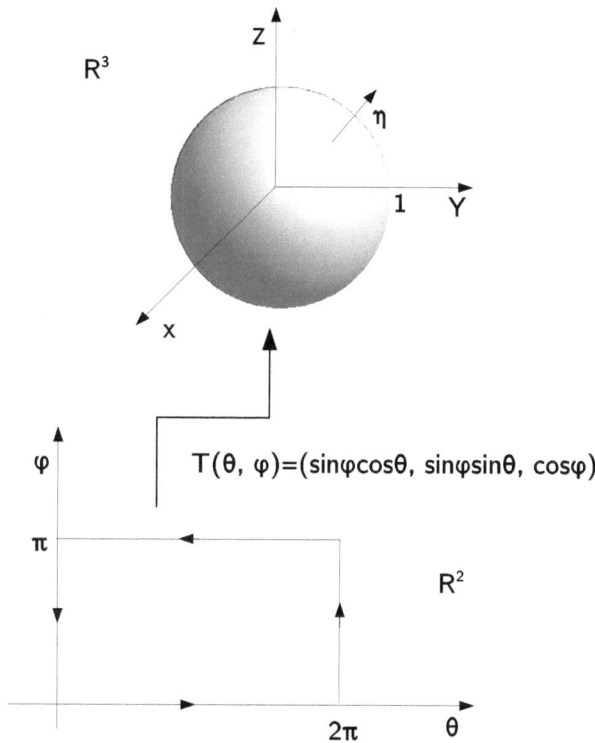

Fig. (7.4). The map transforms the oriented region (θ, ϕ) in the plane into the oriented outer surface of the sphere.

A surface can be oriented indicating with a normal vector $\eta(u,v)$ the orientation of the region. In this sense, a normal vector η that points outwards means that the perimeter of the domain region is traversed counter-clockwise, while a normal vector pointing inwards means that the domain region is traversed clockwise. Two maps are equivalent if they represent the same surface and their normal vectors have the same orientation.

7.3.2. Surface Integral of a Scalar Function

Definition 7.3. A Surface Integral of a Scalar Function (Eq. 7.3) measures the total value of a **scalar magnitude** (Sect. 7.2.3) represented by the real-valued function $f\colon \mathbb{R}^n \to \mathbb{R}$ on a **non-oriented** surface $T\colon \mathbb{R}^2 \to \mathbb{R}^n$.

$$\iint_S f(x,y,z) \circ T(u,v) \, \|T_u \times T_v\| \, dvdu = \int_{u_1}^{u_2} \int_{v_1(u)}^{v_2(u)} f(T(u,v)) \, \|T_u \times$$
$$T_v\| \, dv \, du. \tag{7.3}$$

Example 7.4. As a particular case, if the real-valued function f is equal to 1, then the Surface Integral of Scalar Function represents the **area** of surface T. Let the map $T(\phi,\theta) = (\sin\phi\cos\theta, \sin\phi\sin\theta, \cos\phi)$, $\theta \in [0,2\pi]$, $\phi \in [0,\pi]$ and the real-valued function $f(x,y,z) = 1$. Compute the surface integral of scalar function.

Solution 7.4. $\iint_S \|T_u \times T_v\| dvdu = \int_0^\pi \int_0^{2\pi} \|T_\phi \times T_\theta\| \, d\theta \, d\phi = \int_0^\pi \int_0^{2\pi} \sin\phi \, d\theta \, d\phi =$
$2\pi \int_0^\pi -\sin\phi \, d\phi = 4\pi.$

7.3.3. Surface Integral of a Vector Function

A Surface Integral of a Vector Function measures the effect of a vector field F on an oriented region S, given by its normal vector $\eta(u,v)$ (Eq. 7.4). From the physical point of view, the surface integral of a vector function measures the **work** used over the oriented surface S due to the influence of a vector field.

$$\oiint_S F \circ T(u,v) \cdot \eta(u,v) \, dS = \int_{u_1}^{u_2} \int_{v_1(u)}^{v_2(u)} F(T(u,v)) \cdot \frac{\partial T}{\partial u} \times \frac{\partial T}{\partial v} \, dv \, du. \tag{7.4}$$

Example 7.5. Let the oriented closed surface $S = \{(\theta,\phi) \in \mathbb{R}^2 \mid 0 \le \theta \le 2\pi, 0 \le \phi \le \pi\}$, the vector field $F(x,y,z) = (x,y,z)$, and the map $T(\theta,\phi) = (\sin\phi \cos\theta, \sin\phi \sin\theta, \cos\phi)$ (Fig. **7.4**). (i) Compute the surface integral of the vector function over the vector field. (ii) Compute the same surface integral in the opposite orientation.

Solution 7.5. (i) $\oiint {}_SF \circ T(\theta,\phi) \cdot \eta(\theta,\phi)dS = \int_{\theta_1}^{\theta_2} \int_{\phi_1(\theta)}^{\phi_2(\theta)} F\big(T(\theta,\phi)\big) \cdot \frac{\partial T}{\partial \theta} \times \frac{\partial T}{\partial \phi} \, d\phi$

$d\theta = \int_0^{2\pi} \int_0^{\pi} T(\theta,\phi)) \cdot \frac{\partial T}{\partial \theta} \times \frac{\partial T}{\partial \phi} \, d\phi \, d\theta . \frac{\partial T}{\partial \theta} \times \frac{\partial T}{\partial \phi} =$

$(\sin^2\phi \cos\theta, \sin^2\phi \sin\theta, \sin\phi \cos\phi)$. Then $\int_0^{2\pi} \int_0^{\pi} \sin\phi \, d\phi \, d\theta = -4\pi$. (ii) ${}_SF \circ$

$T(\theta,\phi) \cdot \eta(\theta,\phi) \, dS = \int_{\theta_1}^{\theta_2} \int_{\phi_1(\theta)}^{\phi_2(\theta)} F(T(\theta,\phi)) \cdot \frac{\partial T}{\partial \theta} \times \frac{\partial T}{\partial \phi} \, d\phi \, d\theta = \int_0^{2\pi} \int_0^{\pi} T(\theta,\phi) \cdot$

$\frac{\partial T}{\partial \theta} \times \frac{\partial T}{\partial \phi} \, d\phi \, d\theta .$

$\frac{\partial T}{\partial \theta} \times \frac{\partial T}{\partial \phi} = (-\sin^2\phi\cos\theta, -\sin^2\phi\sin\theta, -\sin\phi\cos\phi)$. Then $\int_0^{2\pi} \int_0^{\pi} -\sin\phi \, d\phi \, d\theta = -4\pi$.

7.4. CASE STUDY: EVALUATING A LINE INTEGRAL

Case 7.1. Let $F(x,y) = (2,3)$. Suppose C is a curve connecting $(0,0)$ to $(1,1)$. Does the value of $\int_C F \cdot dr$ depend on the shape of curve C? If it does not, find the value of the integral. (Case adapted with permission of the author [13]).

We are asked if the value of a line integral is path independent.

Recall:

The line integral $\int_C F \cdot dr$ is path independent if F is conservative. That is, if $F = \nabla\phi$ in some region, then the line integral is unchanged if C is deformed in that region.

$\mathbf{F}(x,y) = u(x,y)\mathbf{i} + v(x,y)\mathbf{j}$ is conservative if, and only if, $\partial_y u(x,y) = \partial_x v(x,y)$.

We identify $u(x,y) = 2$ and $v(x,y) = 3$.

Since $\partial y u = \partial x v = 0$, F is conservative, the value of the line integral does not depend on the shape of the path.

Finding the value of the integral.

Recall:

The fundamental theorem can be used to evaluate a line integral, but $F(x,y)$ has to be conservative.

We have already shown that F is conservative. Hence, we use the fundamental theorem of line integrals:

If $F(x, y) = i\nabla\phi(x, y)$ and C is any curve going from (x_0, y_0) to (x_1, y_1)

$$\int_C \mathbf{F} \cdot d\mathbf{r} = \phi(x_1, y_1) - \phi(x_0, y_0).$$

To apply this theorem, we need to find the potential function $\phi(x, y)$.

We can derive that the potential function for $F(x, y)$ is $\phi(x, y) = 2x + 3y$.

Applying the fundamental theorem of line integrals, we can compute

$$\int_C \mathbf{F} \cdot d\mathbf{r} = \phi(1,1) - \phi(0,0) = 5.$$

7.5. CASE STUDY: SURFACE AREA OF A TORUS

Case 7.2. Parameterise the torus T_{ab} in hand (with a circle of radius b, whose centre is dragged round a circle of radius a) (Fig. **7.5**). (Case adapted with permission of the author [14]).

$$T(u, v) = (T_1(u, v), T_2(u, v), T_3(u, v)) \quad \text{for} 0 \le u, v \le 2\pi,$$

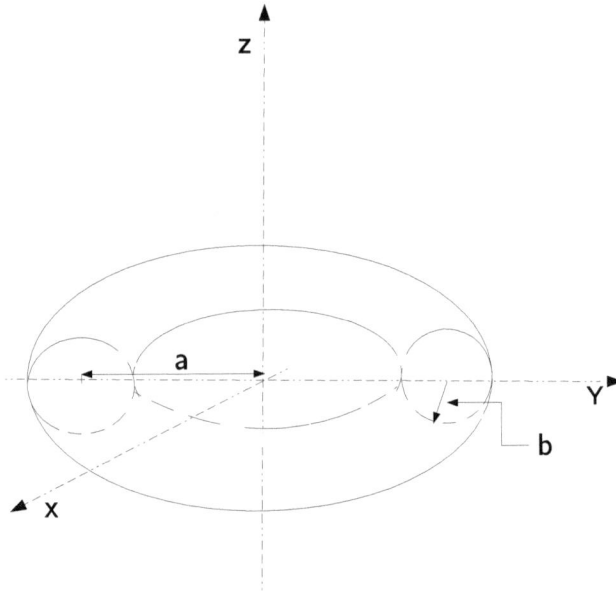

Fig. (7.5). Geometric representation of a Torus.

We can compute the stretch factor (Jacobian) for the surface area of the torus and integrate it.

Note 7.5. Note that we need to have $a > b$, so the torus does not "turn inside out". This means that when we need to know what $|a + b\cos v|$ is, we have to calculate $a + b\cos v \cdots$

We compute

$$T_u \quad = \frac{\partial T}{\partial u} = (-(a + b\cos v)\sin u, (a + b\cos v)\cos u, 0)$$

$$T_v \quad = \frac{\partial T}{\partial v} = (-b\sin v\cos u, -b\sin v\sin u, b\cos v)$$

Now we compute the normal vector $T_u \times T_v$.

$$T_u \times T_v \quad =$$
$$([a + b\cos v\cos u][b\cos v] - 0, -([-(a + b\cos v)\sin u][b\cos v] - 0),$$
$$[-(a + b\cos v)\sin u][-b\sin v\sin u] - [(a + b\cos v)\cos u][-b\sin v\cos u])$$
$$= (a + b\cos v)(b\cos u\cos v, b\sin u\cos v, b\sin^2 u\sin v + b\cos^2 u\sin v)$$
$$= (a + b\cos v)(b\cos u\cos v, b\sin u\cos v, b\sin v)$$
$$= b(a + b\cos v)(\cos u\cos v, \sin u\cos v, \sin v)$$

The vector on the right end ought to look familiar; that is how we denote points on the unit sphere in spherical coordinates. Then

$$||T_u \times T_v||^2 \quad = b^2(a + b\cos v)^2(\cos^2 u\cos^2 v + \sin^2 u\cos^2 v + \sin^2 v)$$
$$= b^2(a + b\cos v)^2(\cos^2 v + \sin^2 v)$$
$$= b^2(a + b\cos v)^2$$

So, $||T_u \times T_v|| = |b(a + b\cos v)| = ab + b^2\cos v$. This is the term we now integrate over the domain of our parameterised surface $0 \leq u, v \leq 2\pi$, which gives

$$\text{Area}(T_{ab}) = \int_0^{2\pi} \int_0^{2\pi} ab + b^2\cos v \, dv \, du$$
$$= \int_0^{2\pi} abv + b^2\sin v|_0^{2\pi} \, du$$
$$= \int_0^{2\pi} (2\pi ab + 0) - (0 + 0) \, du$$
$$= \int_0^{2\pi} 2\pi ab \, du = (2\pi ab)(2\pi)$$
$$= 4\pi^2 ab.$$

7.6. EXERCISES

Exercise 7.1. Consider a segment of length 2π in $\mathbb{R}$. (i) Define a map from $\mathbb{R}$ to $\mathbb{R}^2$ that transforms this line into the ellipse $\dfrac{x^2}{a^2} + \dfrac{y^2}{b^2} = 1$ in the plane. (ii) Draw the figure.

Exercise 7.2. Consider a segment of length 2π in $\mathbb{R}$. (i) Define a map from $\mathbb{R}$ to $\mathbb{R}^3$ that transforms this line into a unit circle projected onto the plane $f(x,y) = 1 - x - y$. (ii) Draw the figure.

Exercise 7.3. Consider a unit circle in $\mathbb{R}^2$. (i) Define a map from $\mathbb{R}^2$ to $\mathbb{R}^3$ that transforms the unit circle into the unit upper sphere. (ii) Draw the figure.

Exercise 7.4. Consider a unit circle in $\mathbb{R}^2$ with centre at the origin (Example 6) and the vector field $F(x,y) = (-y, x)$. (i) Compute the line integral of vector field in counter-clockwise direction. (ii) Compute the line integral of scalar field in clockwise direction.

Exercise 7.5. Consider a unit circle in the plane $f(x,y) = 1$ with centre at the origin, and the vector field $F(x,y,z) = (-y, x, 1)$. (i) Compute the line integral of vector field in counter-clockwise direction. (ii) Compute the line integral of vector field in clockwise direction.

Exercise 7.6. Consider a unit upper sphere and the vector field $F(x,y,z) = (x,y,z)$. (i) Compute the surface integral of vector field in counter-clockwise direction. (ii) Compute the line integral of vector field in clockwise direction.

Exercise 7.7. Consider the trajectory $c(t) = (t, t^2), t \in [0, 2\pi]$ and the real valued function $f(x,y) = x^2 + y^2$. (i) Compute the line integral over this trajectory. (ii) Draw the figure.

Exercise 7.8. Consider the map $T(u,v) = (u, 2 + v, 3uv)$, where $u \in [0,1]$ and $v \in [0,1]$. Compute the area of the parameterized surface.

Part IV

FUNDAMENTAL THEOREMS

CHAPTER 8

Theorems of Vector Calculus

Abstract: This chapter reviews the theorems of Vector Calculus: **Green's theorem**, **Stokes' theorem**, and **Gauss' theorem**. The approach will be done using mappings, to show the operative simplification that can be obtained.

Keywords: Closed counter-clockwise orientation boundary, Counter-clockwise direction, Double integral, Bounded region, Gauss' theorem, Green's theorem, integral theorems, Maps, Normal vector, Parameterization of a curve, Stokes' theorem, Triple integral, Vector valued-function, Volume bounded.

8.1. INTEGRAL THEOREMS

8.1.1. Green's Theorem on $\mathbb{R}^2$

Definition 8.1. Let $D \subset \mathbb{R}^2$ be a region and ∂D be its closed counter-clockwise orientation boundary. Let the vector-valued function $F \colon \mathbb{R}^2 \to \mathbb{R}^2$ be on the D region and the **unit vector** k [25]. Then

$$\oint_{\partial D} F \circ c(t) \cdot c'(t)\, ds = \iint_D (\nabla \times F) \cdot \mathrm{k}\, dA.$$

Green's theorem states that the effect of the vector-valued function F over the close oriented curve ∂D, counter-clockwise direction, is equivalent to the volume bounded by the region D and the graph of the real-valued function $(\nabla \times F) \cdot \mathrm{k}$.

Example 8.1. Let the vector-valued function $F(x,y) = (x^2, y^3)$, the region D bounded by $x \geq 0, y \geq 0, y = x^3$, and $y = x$. (i) Verify Green's theorem. (ii) Draw regions D and ∂D.

Solution 8.1. (i) $\oint_{\partial D} F \cdot ds = \oint_{C_1} F(t) \circ c_1(t) \cdot c_{1'}(t) dt + \oint_{C_2} F(t) \circ c_2(t) \cdot c_{2'}(t)\, dt$, where the curves $c_1(t) = (t, t^3), t \in [0,1]$, and $c_2(t) = (t,t), t \in [1,0]$ bound region D. $\int_0^1 (t^2, t^9) \cdot (1, 3t^2)\, dt = \int_0^1 t^2 + 3t^{11}\, dt = \frac{7}{12}$. $\int_1^0 (t^2, t^3) \cdot (1,1)\, dt = -\int_0^1 t^2 + t^3\, dt = -\frac{7}{12}$. $\oint_{\partial D} F \cdot ds = 0$. $\iint_D (\nabla \times F) \cdot \mathrm{k}\, dA =$

$\int_0^1 \int_{x^3}^x (0,0,0) \cdot (0,0,1) \, dydx = 0$. The double integral can be solved with a map from $T \colon \mathbb{R}^2 \to \mathbb{R}^2$ (Eq. 5.6). (ii) See Fig. **(8.1)**.

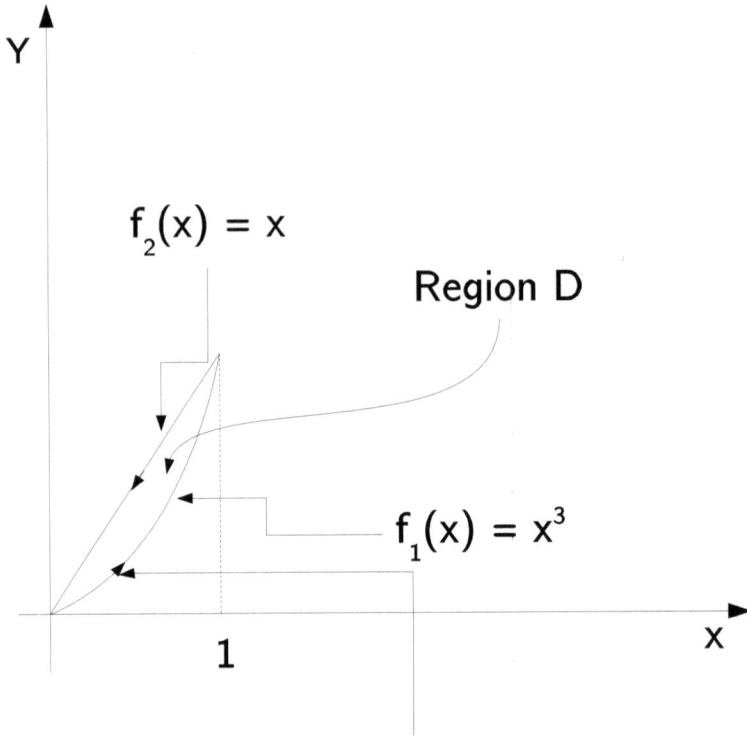

Curves   D in counter-clockwise orientation

Fig. (8.1). The region D and the curves ∂D.

Example 8.2. Let the vector-valued function $F(x,y) = (-y,x)$ and the region D bounded by $\dfrac{x^2}{9} + \dfrac{y^2}{4} = 1$. (i) Verify Green's theorem. (ii) Draw regions D and ∂D.

Solution 8.2. (i) $\oint_{\partial D} F \cdot ds = \oint_c F(t) \circ c(t) \cdot c'(t) \, dt$, where $c(t) = (3\cos t, 2\sin t), t \in [0, 2\pi]$ bound region D. $\int_0^{2\pi} (-2\sin t, 3\cos t) \cdot (-3\sin t, 2\cos t) \, dt = \int_0^{2\pi} 6\sin^2 t +$

$6\cos^2 t \, dt = 12\pi$. $\iint_D (\nabla \times F) \cdot k \, dA = 4 \int_0^3 \int_{-\sqrt{4-\frac{4}{9}x^2}}^{\sqrt{4-\frac{4}{9}x^2}} (0,0,2) \cdot (0,0,1) \, dydx = 4$

$(2)(\frac{6}{4}\pi) = 12\pi$. The double integral can be solved with a map from $T \colon \mathbb{R}^2 \to \mathbb{R}^2$ (Eq. 5.6). (ii) See Fig. **(8.2)**.

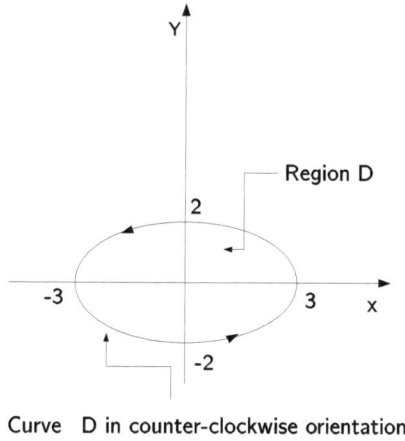

Curve D in counter-clockwise orientation

Fig. (8.2). The region D and the curve ∂D.

8.1.2. Green's Theorem Case

Definition 8.2. Let $D \subset \mathbb{R}^2$ be a region and ∂D be its closed counter-clockwise orientation boundary. Let the vector-valued function $F: \mathbb{R}^2 \to \mathbb{R}^2$ be on the D region and the **normal unit vector η** be a normal vector to the outer region D [25]. Then

$$\oint_{\partial D} F \circ c(t) \cdot \eta(t) \, ds = \iint_D (\nabla \times F) \, dA$$

where $\eta(t)$ is given by

$$\eta(t) = \frac{1}{\sqrt{x\prime(t)^2 + y\prime(t)^2}} (y'(t), -x'(t))$$

This case introduces an important variant to the Green's theorem. The line integral, at the left side of the equality, is not a proper line integral because vector $\eta(t)$ is included in its integrand; which is normal to the first derivative $c'(t)$ of the parameterization of the curve ∂D.

Example 8.3. Let the vector-valued function $F(x, y) = (x^2, y^3)$, the region D bounded by $x \geq 0, y \geq 0, y = x^3$, and $y = x$ (Ex. 1.1). (i) Verify Green's theorem. (ii) Draw regions D and ∂D.

Solution 8.3. (i) $\oint_{\partial D} F \circ c(t) \cdot \eta(t) ds = \oint_{C_1} F \times c_1(t) \cdot \eta_1(t) \, dt + \oint_{C_2} F \times c_2(t) \cdot \eta_2$

$(t)\, dt$, where the curves $c_1(t) = (t, t^3), t \in [0,1]$ and $c_2(t) = (t, t), t \in [1,0]$ bound region D, and their normal unit vectors are given by $\eta_1(t) = (3t^2, -1)$ and $\eta_2(t) = (1, -1)$. $\int_0^1 (t^2, t^9) \cdot (3t^2, -1)\, dt = \int_0^1 3t^4 - t^9\, dt = \frac{1}{2}. -\int_1^0 (t^2, t^3) \cdot (1, -1)\, dt = \int_0^1 t^2 - t^3\, dt = -\frac{1}{12}.$ $\oint_{\partial D} F \cdot ds = \frac{5}{12}.$ $\iint_D (\nabla \times F)\, dA = \int_0^1 \int_{x^3}^x 2x + 3y^2\, dydx = \frac{5}{12}.$

The double integral can be solved with a map from $T \colon \mathbb{R}^2 \to \mathbb{R}^2$ (Eq. 5.6). (ii) See Fig. (**8.1**).

Example 8.4. Let the vector-value function $F(x, y) = (-y, x)$ and the region D bounded by $\frac{x^2}{9} + \frac{y^2}{4} = 1$. (i) Verify Green's theorem. (ii) Draw regions D and ∂D.

Solution 8.4. (i) $\oint_{\partial D} F \cdot ds = \oint_c F \times c(t) \cdot \eta(t)\, dt$, where $c(t) = (3\cos t, 2\sin t), t \in [0, 2\pi]$ bound region D and the vector $\eta(t) = \frac{1}{\sqrt{9\cos^2 t + 4\sin^2 t}}(2\cos t, -3\sin t)$. $\int_0^{2\pi} (2\sin t, 3\cos t) \cdot \frac{1}{\sqrt{9\cos^2 t + 4\sin^2 t}}(2\cos t, -3\sin t)\, dt = \int_0^{2\pi} \frac{-13\sin t\cos t}{\sqrt{9\cos^2 t + 4\sin^2 t}}\, dt = 0.$ $\iint_D (\nabla \circ F)\, dA = 0 \int_0^3 \int_{-\sqrt{4 - \frac{4}{9}x^2}}^{\sqrt{4 - \frac{4}{9}x^2}} dydx = 0.$ The double integral can be solved with a map from $T \colon \mathbb{R}^2 \to \mathbb{R}^2$ (Eq. 5.6). (ii) See Fig. (**8.2**).

8.1.3. Stokes' Theorem in $\mathbb{R}^3$

Definition 8.3. Let $D \subset \mathbb{R}^3$ be a closed region and ∂D be its counter-clockwise orientation surface. Let the vector-valued function $F \colon \mathbb{R}^3 \to \mathbb{R}^3$ be on the D region and the **normal vector** $T_v \times T_u$ [25] be perpendicular to the surface ∂D.

$$\oint_{\partial D} F \circ c(t) \cdot c'(t)\, ds = \iint_S (\nabla \times F) \cdot T_v \times T_u\, dv\, du.$$

Stokes' theorem states that the effect of the vector-valued function F over the close oriented curve ∂D, counter-clockwise direction, is equivalent to the volume bounded by the region D and the graph of the real-valued function $(\nabla \times F) \cdot T_v \times T_u$.

Example 8.5. Let the vector-value function $F(x, y, z) = (x^2, y^2, z^2)$, the region D bounded by the cylinder surface $x^2 + y^2 = 1$, and the real-valued function $f(x, y) = 1 - x - y$. (i) Verify Stokes' theorem. (ii) Draw regions D and ∂D.

Solution 8.5. (i) $\oint_{\partial D} F(c(t)) \cdot c'(t) dt$, where the curve $c(t)$ is $c(t) = (\cos t, \sin t, 1 - \cos t - \sin t), t \in [0, 2\pi]$. $\int_0^{2\pi} (\cos^2 t, \sin^2 t, (1 - \cos t - \sin t)^2) \cdot (-\sin t, \cos t, \sin t - \cos t) \, dt = \int_0^{2\pi} -\cos^2 t \sin t + \sin^2 t \cos t + (1 - \cos t - \sin t)^2 (\sin t - \cos t) \, dt = 0$. On the other hand $\iint_D \nabla \times F(T(r, \theta)) \cdot T_r \times T_\theta \, d\theta \, dr$, with $T(r, \theta) = (r\cos\theta, r\sin\theta, 1 - \cos\theta - \sin\theta)$. $\int_0^1 \int_0^{2\pi} (0,0,0) \cdot (r\sin\theta - r\cos\theta, -\cos\theta - \sin\theta, \sin\theta + r\sin\theta) d\theta \, dr = 0$ (see Eq. 5.6).
(ii) See Fig. (**8.3**).

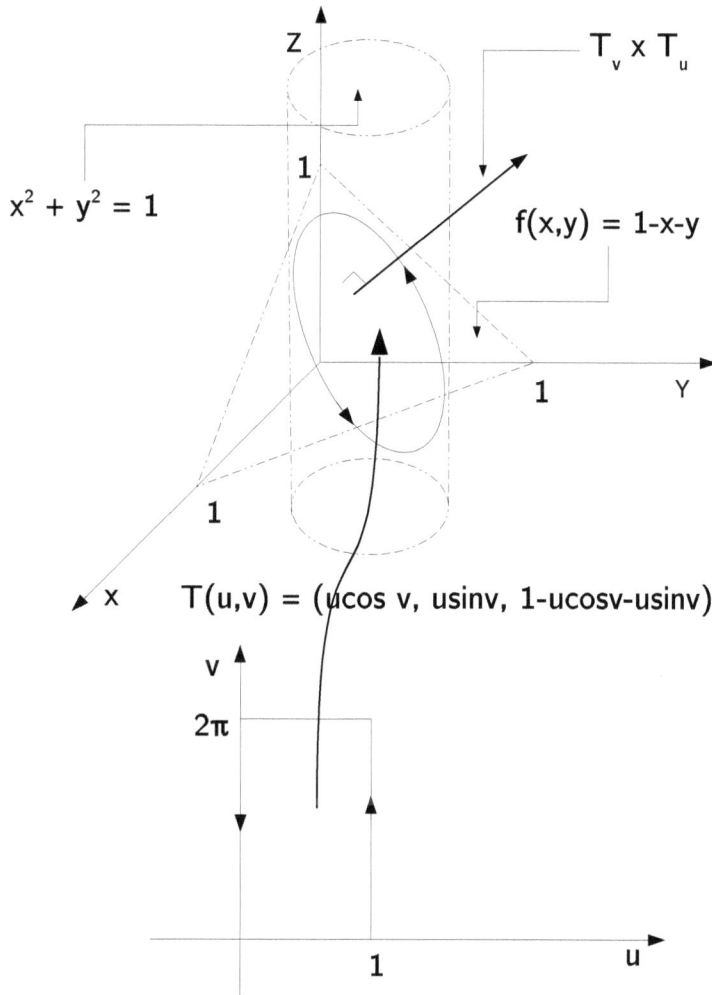

Fig. (8.3). Geometrical description of map T from $\mathbb{R}^2 \to \mathbb{R}^3$.

8.1.4. Gauss' Theorem in $\mathbb{R}^3$

Definition 8.4. Let $W \subset \mathbb{R}^3$ be a closed and solid region and ∂W be its counter-clockwise orientation surface. Let the vector-valued function $F: \mathbb{R}^3 \to \mathbb{R}^3$ be on the W region and the **normal vector** $T_v \times T_u$ [25] be perpendicular to surface W.

$$\oiint_{\partial W} F \circ T(u,v) \cdot T_v \times T_u \, dv \, du = \iiint_W (\nabla \cdot F) \, dz \, dy \, dx.$$

Gauss' theorem states that the effect of the vector-valued function F over the close and oriented surface ∂W, counter-clockwise direction, is equivalent to the volume bounded by the region W and the graph of the real-valued function $\nabla \cdot F$.

Example 8.6. Let the vector-value function $F(x,y,z) = (x^2, y^2, z^2)$, the region W bounded by the solid cylinder $x^2 + y^2 = 1$, and the real-valued function $f(x,y) = 1 - x - y$. (i) Verify Gauss' theorem. (ii) Draw regions W and ∂W.

Solution 8.6. (i) $\oint_{\partial W} F(c(t)) \cdot c'(t) dt$, where the curve $c(t)$ is $c(t) = (\cos t, \sin t, 1 - \cos t - \sin t), t \in [0, 2\pi]$. $\int_0^{2\pi} (\cos^2 t, \sin^2 t, (1 - \cos t - \sin t)^2) \cdot (-\sin t, \cos t, \sin t - \cos t) \, dt = \int_0^{2\pi} \cos^2 t \sin t + \sin^2 t \cos t + (1 - \cos t - \sin t)^2 (\sin t - \cos t) \, dt = 0$. On the other hand $\iint_W \nabla \times F(T(r, \theta)) \cdot T_r \times T_\theta \, d\theta \, dr$, with $T(r, \theta) = (r\cos\theta, r\sin\theta, 1 - \cos\theta - \sin\theta)$. $\int_0^1 \int_0^{2\pi} (0,0,0) \cdot (r\sin\theta - r\cos\theta, -\cos\theta - \sin\theta, \sin\theta + r\sin\theta) \, d\theta$

$dr = 0$ (see Eq. 5.6). (ii) See Fig. (**8.4**).

8.2. Case Study: $\oint_C (-y + \frac{1}{3}y^3 + x^2 y) dx$

Case 8.1. Let C be a circle of radius a centered at the origin, traversed counter-clockwise, where the non-zero value of a is

$$\oint_C (-y + \frac{1}{3}y^3 + x^2 y) dx = 0.$$

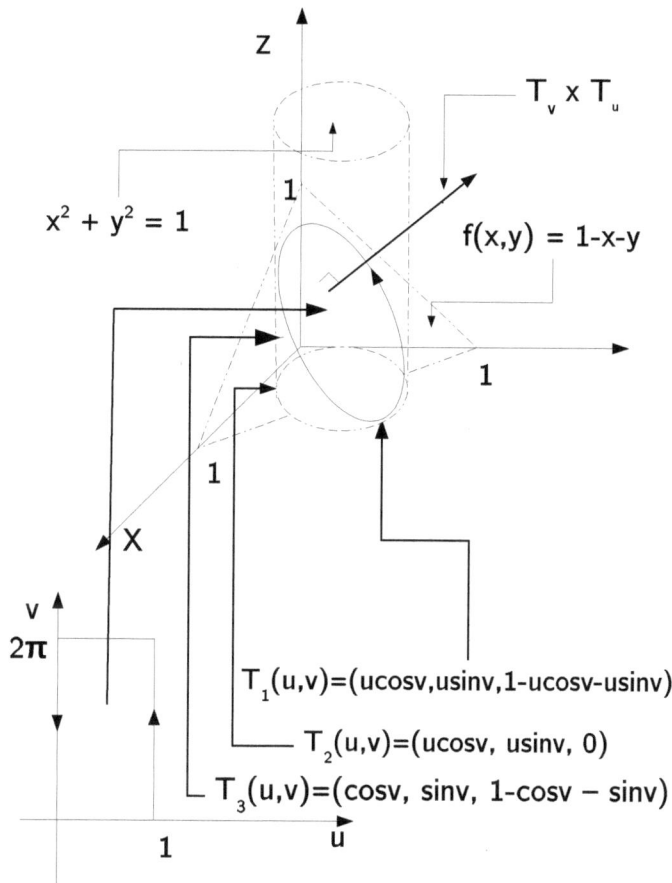

Fig. (8.4). Geometrical description of map T from $\mathbb{R}^2 \to \mathbb{R}^3$.

Case reproduced with the permission of the author [15]

Recall:

There are several ways to compute a line integral $\int_C \mathbf{F}(x,y) \cdot d\mathbf{r}$:

Direct parameterisation. The fall-back method is convenient when C has a parameterisation that simplifies $F(x,y)$.

The theorem of line integrals is usually the easiest to use, but $F(x,y)$ has to be conservative.

Green's theorem only applies when C is closed and F is well-behaved inside C.

Selecting an integration method. Our initial approach to solving the line integral is by direct parameterisation The parameterisation of a circle of radius a is

$$x(\theta) = a\cos\theta, \quad y(\theta) = a\sin\theta.$$

Hence,

$$dx = -a\sin\theta d\theta, \quad dy = a\cos\theta d\theta.$$

The integral gives rise to terms like $x^2 y \, dx = -a^4 \cos^2\theta \, \sin^2\theta \, d\theta$. Terms like this can be integrated but they can be messy; we will try to find an easier method of integration. Our second approach is to solve the integral using the theorem of line integrals.

Recall:

The theorem applies if the vector field $F(x,y)$ is conservative. This is when $F(x,y) = \nabla\phi(x,y)$ for some ϕ.

$\mathbf{F}(x,y) = u(x,y)\mathbf{i} + v(x,y)\mathbf{j}$ is conservative if, and only if,

$$\partial_y u(x,y) = \partial_x v(x,y).$$

We identify $u(x,y) = -y + \frac{1}{3}y^3 + x^2 y$ and $v(x,y) = 0$. Since $\partial_y u \neq \partial_x v$, the vector field, is not conservative we cannot use the theorem. Our third approach is to solve the integral using Green's theorem.

Solution using Green's Theorem. Green's theorem only applies when C is closed and F is well-behaved inside C. We note that C is closed because it is a circle. The vector field is a polynomial in x, y, so it is well behaved inside C. If $F = L(x,y)\mathbf{i} + M(x,y)\mathbf{j}$ is differentiable inside a closed and positively oriented curve C, then

$$\oint_C L(x,y)dx + M(x,y)dy = \iint_D (\partial_x M(x,y) - \partial_y L(x,y))dxdy,$$

where D is the region inside C. We identify $L(x,y) = -y + \frac{1}{3}y^3 + x^2 y$ and $M(x,y) = 0$ We compute the partial derivatives

$$\partial_y L(x,y) = -1 + y^2 + x^2 \tag{8.1}$$

$$\partial_x M(x, y) = 0 \tag{8.2}$$

Applying the theorem and computing the partial derivatives, we see that

$$\oint_C (-y + \frac{1}{3}y^3 + x^2 y)dx = \iint_D (1 - y^2 - x^2)dxdy$$

where D is the disc of radius a. We can compute

$$\iint_D (1 - y^2 - x^2)dxdy = \frac{\pi}{2}(2a^2 - a^4)$$

We want to find the value of a such that

$$\frac{\pi}{2}(2a^2 - a^4) = 0$$

Dividing it by the constant and a^2, we get $2 = a^2$. The radius a by which the line integral is zero is $a = \sqrt{2}$.

8.3. CASE STUDY: $\oint_C xy^2 dx + x^2 y dy + y dz$

Case 8.2. We will use Gauss' theorem to evaluate the surface integral of $F(x, y, z) = (xy^2, x^2 y, y)$, through a surface S of the cylinder W bounded by $x^2 + y^2 = 1$ and the planes $z = 2$ y $z = -2$. (Case taken from [16]).

The instructions of this case suggest the resolution of the triple integral of the divergence of F, instead of directly calculating the net flux of the field, which in this case would be given by 3 surface integrals: one for the lateral surface of the cylinder and two for the top and bottom surfaces of it. Both the vector field and the regions, the interior region of the cylinder and the boundary regions, meet the conditions of the theorem. Note that the statement does not mention if the flux is inward or outward, so we will take it as outward. (i) Draw the solid and the boundary regions with a normal vector for each of the 3 parts forming it. (ii) Evaluate the surface integral.

(i) Fig. **(8.5)**

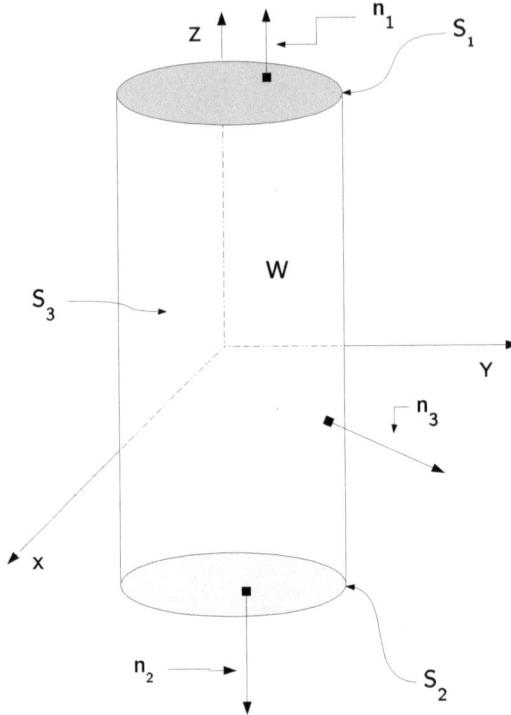

Fig. (8.5). Geometrical description of the surface S and the region W.

(ii) We will need the divergence of the vector field

$$\nabla \cdot F = \frac{\partial}{\partial x}xy^2 + \frac{\partial}{\partial y}x^2y + \frac{\partial}{\partial z}y = y^2 + x^2.$$

The surface integral on the outside is equivalent, by the Gauss' theorem, to the triple integral

$$\oiint_{\partial S = S_1 \cup S_2 \cup S_3} F \circ T(u,v) \cdot T_v \times T_u \, dv \, du$$

$$= \iiint_W (\nabla \cdot F) \, dz \, dy \, dx$$

$$= \iiint_W y^2 + x^2 \, dz \, dy \, dx$$

$$= \int_{-2}^{2} \int_{0}^{2\pi} \int_{0}^{1} r^3 dr d\theta dz$$

$$= 2\pi. \qquad \qquad (8.3)$$

8.4. EXERCISES

Exercise 8.1. Use the Green' theorem to calculate $\oint_C (2y + \sqrt{1 + x^5})dx + (5x - e^{y^2})dy$, where $C := x^2 + y^2 \leq 4$ [49].

Exercise 8.2. Use the Green' theorem to calculate $\oint_C y^2 dx + 3xy dy$, where C is the counter-clockwise direction oriented boundary of the upper-half of unit disc a [50].

Exercise 8.3. Let the vector-value function $F(x, y, z) = (y, z, x)$ [51], the region S (positive orientation) be bounded by the sphere $x^2 + y^2 + z^2 = 1$, and the real-valued function $f(x, y) = -x - y$. (i) Compute the Stokes' line integral. (ii) Draw regions S and ∂S.

Exercise 8.4. Explain the Maxwell-Faraday equation according to the Stokes' theorem.

Exercise 8.5. Evaluate the integral $\iint_S (3xi + 2yj)\, dA$, where S is the solid sphere $x^2 + y^2 + z^2 \leq 16$ [52], according to the Guass' theorem.

Exercise 8.6. Use the Gauss' theorem to calculate $\oiint_S F\, ds$, where S is the lateral surface of box B with vertices $(\pm 1, \pm 2, \pm 3)$, and the normal vector $F(x, y, z) = (x^2 z^3, 2xyz^3, xz^4)$ [53] is pointing outwards.

Part V

GEOMETRIC PRODUCT

CHAPTER 9

Geometric Algebra

Abstract: This chapter intends to be a survey on Exterior Algebra. This algebra is attributed to Hermann Grassmann [Die lineare Ausdehnungslehre, ein neuer Zweig der Mathematik, 1842], and it is formed by two operators: the **exterior product** and **inner product**. William K. Clifford unified both products under the **geometric product** operator and David Hestenes improved the geometric and computer aspects of Geometric algebra. Here, we review the main operators and their properties, as well as their application in the representation of lines and planes.

Keywords: Bivector, Differential forms, Differentiation, Dot product, Dual space, Exterior product, Geometric algebra, Geometric product, Grassmann algebra, Inner product, Norm, Oriented plane, Orthogonal basis, Reflections, Rigid body motions, Rotations.

9.1. GEOMETRIC ALGEBRA $\mathbb{G}_2$ in $\mathbb{R}^2$

Definition 9.1. We define the unitary associative algebra named **Geometric Algebra** [54] $\mathbb{G}_2 = \mathbb{G}_2(\mathbb{R}^2)$ and four elements: scalars, σ_1, σ_2 vectors, $\sigma_1 \wedge \sigma_2$, **bivectors** (or **equivalently** $\sigma_1 \sigma_2$), which form the **orthonormal** basis $(1, \sigma_1, \sigma_2, \sigma_1 \wedge \sigma_2)$ that meets $\sigma_i \wedge \sigma_i = 1$ and $\sigma_i \wedge \sigma_j = -\sigma_j \wedge \sigma_i$. For the algebra $\mathbb{G}_2$ there is an arbitrary element

$$v = v_0 + v_1 \sigma_1 + v_2 \sigma_2 + v_{12} \sigma_1 \wedge \sigma_2 \in \mathbb{G}_2.$$

From these two elements $a = a_0 + a_1 \sigma_1 + a_2 \sigma_2 + a_{12} \sigma_1 \sigma_2$, and $b = b_0 + b_1 \sigma_1 + b_2 \sigma_2 + b_{12} \sigma_1 \sigma_2 \in \mathbb{G}_2$, the **geometric product** (Eq. 9.1) is defined as:

$$ab = (a_0 + a_1 \sigma_1 + a_2 \sigma_2 + a_{12} \sigma_1 \sigma_2)(b_0 + b_1 \sigma_1 + b_2 \sigma_2 + b_{12} \sigma_1 \sigma_2) \tag{9.1}$$

Alternatively, the geometric product is defined by $ab = a \cdot b + a \wedge b$, where the term $\mathbf{a} \cdot \mathbf{b}$ is called **inner product** (Def. 9.3) and the term $\mathbf{a} \wedge \mathbf{b}$ is called **exterior product** (Def. 9.2).

Note 9.1. The expression of the exterior product $\sigma_1\sigma_2$ is only used with the elements σ_i of the orthogonal basis, this expression will **never** be used with elements of $\mathbb{G}_2$, *i.e. $a \wedge b$.*

Note 9.2. The geometrical meaning of the **bivector** $\sigma_1 \wedge \sigma_2$ is the **oriented** area spanned by the vectors σ_1 and σ_2, whose value is 1 (Fig. **9.1**).

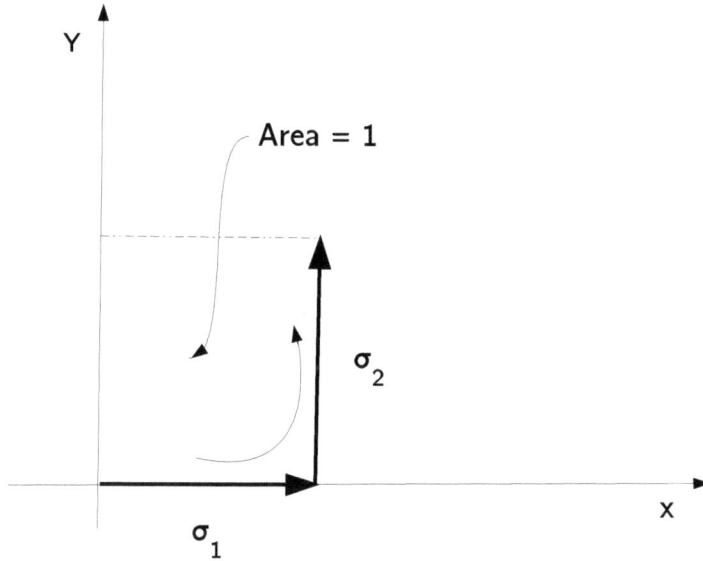

Fig. (9.1). The oriented plane formed by the vectors σ_1 and σ_2.

Example 9.1. Let two elements $a = (1,-2)$ and $b = (2,3) \in \mathbb{G}_2$. (i) Express these elements using the orthonormal basis. (ii) Obtain the geometric product ab. (iii) Obtain the geometric product ba.

Solution 9.1. (i) $a = \sigma_1 - 2\sigma_2$, and $b = 2\sigma_1 + 3\sigma_2$. (ii) $ab = (\sigma_1 - 2\sigma_2)(2\sigma_1 + 3\sigma_2) = 2\sigma_1\sigma_1 - 6\sigma_2\sigma_2 + 3\sigma_1\sigma_2 - 4\sigma_2\sigma_1 = -4 + 3\sigma_1\sigma_2 + 4\sigma_1\sigma_2 = -4 + 7\sigma_1\sigma_2$. (iii) $(2\sigma_1 + 3\sigma_2)(\sigma_1 - 2\sigma2) = (2-6) - 4\sigma_1\sigma_2 - 3\sigma_1\sigma_2 = -4 - 7\sigma_1\sigma_2$.

9.1.1. Exterior Product: $a \wedge b$

Definition 9.2. For two elements $a = a_0 + a_1\sigma_1 + a_2\sigma_2 + a_{12}\sigma_1 \wedge \sigma_2$ and $b = b_0 + b_1\sigma_1 + b_2\sigma_2 + b_{12}\sigma_1 \wedge \sigma_2 \in \mathbb{G}_2$, we define

$$a \wedge b = \tfrac{1}{2}(ab - ba).$$

Example 9.2. Let two elements $a = (1, -2)$ and $b = (2, 3) \in \mathbb{G}_2$. Obtain the exterior product $a \wedge b = \frac{1}{2}(ab - ba)$. From Ex. 9.1 $a \wedge b = \frac{1}{2}(ab - ba) = 7\sigma_1\sigma_2$.

Example 9.3. Consider two elements i and $j \in \mathbb{G}_2$. (i) Express these elements using the orthonormal basis. (ii) With the elements ij and ji obtain the exterior product $i \wedge j$. (iii) Obtain the exterior product with the elements i and $\alpha i, \alpha \in \mathbb{G}_2$.

Solution 9.3. (i) $i = \sigma_1$ and $j = \sigma_2$. (ii) $ij = \sigma_1\sigma_2$, $ji = -\sigma_1\sigma_2$, then $i \wedge j = \sigma_1\sigma_2$. (iii) $i(\alpha i) = \sigma_1(\alpha\sigma_1) = \alpha\sigma_1\sigma_1 = \alpha$. $(\alpha i)i = \alpha\sigma_1\sigma_1 = \alpha$, then $i \wedge \alpha i = 0$.

9.1.2. Inner Product: $a \cdot b$

Definition 9.3. For two elements $a = a_0 + a_1\sigma_1 + a_2\sigma_2 + a_{12}\sigma_1\sigma_2$ and $b = b_0 + b_1\sigma_1 + b_2\sigma_2 + b_{12}\sigma_1\sigma_2 \in \mathbb{G}_2$, we define

$$a \cdot b = \frac{1}{2}(ab + ba).$$

Example 9.4. Consider elements $a = b = \sigma_1 + \sigma_2 \in \mathbb{G}_2$. (i) Obtain the geometric products ab and ba. (ii) Determine $a \cdot b$. (iii) Determine $a \wedge b$.

Solution 9.4. (i) $ab = ba = \sigma_1\sigma_1 + \sigma_2\sigma_2 + \sigma_1\sigma_2 - \sigma_1\sigma_2 = 2$. (ii) $a \cdot b = 2$. (iii) $a \wedge b = 0$.

Example 9.5. Let two elements $a = \sigma_1\sigma_2$, and $b = \sigma_2 \in \mathbb{G}_2$. (i) Obtain the geometric products ab and ba. (ii) From Def. 9.3, determine $a \cdot b$. (iii) From Def. 9.2, determine $a \wedge b$.

Solution 9.5. (i) $ab = \sigma_1\sigma_2\sigma_2 = \sigma_1$. $ba = -\sigma_1$. (ii) $a \cdot b = 0$. (iii) $a \wedge b = \sigma_1$.

9.1.3. Distributivity: $a(b + c)$

Definition 9.4. For three elements a, b, and $c \in \mathbb{G}_2$, then

$$a(b + c) = ab + ac$$

Proof.

$$ab + ac = a \cdot b + a \wedge b + a \cdot c + a \wedge c$$
$$= a \cdot b + a \cdot c + a \wedge b + a \wedge c$$
$$= a \cdot (b + c) + a \wedge (b + c) \tag{9.2}$$
$$= a(b + c)$$

Example 9.6. Let three elements $a = \sigma_1, b = \sigma_2,$ and $c = \sigma_1 + \sigma_2$. (i) Determine $a(b + c)$. (ii) Determine $ab + ac$. (iii) Is the distributivity fulfilled?

Solution 9.6. (i) $a(b + c) = \sigma_1(\sigma_2 + \sigma_1 + \sigma_2) = 1 + 2\sigma_1\sigma_2$. (ii) $ab = \sigma_1\sigma_2$ and $ac = \sigma_1(\sigma_1 + \sigma_2) = 1 + \sigma_1\sigma_2$, then $ab + ac = 1 + 2\sigma_1\sigma_2$. (iii) From (i) and (ii) yes, it is.

9.1.4. Distributivity: $a \wedge (b + c)$

Definition 9.5. For three elements a, b, and $c \in \mathbb{G}_2$, then

$$a \wedge (b + c) = a \wedge b + a \wedge c$$

Proof.

$$a \wedge (b + c) = -ab - ac + a \cdot (b + c)$$
$$= -ab - ac + a \cdot b + a \cdot c$$
$$= (-ab + a \cdot b) + (-ac + a \cdot c) \tag{9.3}$$
$$= a \wedge b + a \wedge c$$

Example 9.7. Let three elements $a = 2, b = 3\sigma_1 - 2\sigma_2,$ and $c = \sigma_1 + \sigma_2$. (i) Determine $a(b + c)$ and $(b + c)a$. (ii) Determine $a \wedge (b + c)$. (iii) Determine $a \wedge b$ and $a \wedge c$. (iv) Is the distributivity over the exterior product fulfilled?

Solution 9.7. (i) $a(b + c) = (b + c)a = 8\sigma_1 - 2\sigma_2$. (ii) $a \wedge (b + c) = 0$. (iii) $a \wedge b = 0$ and $a \wedge c = 0$, then $a \wedge b + a \wedge c = 0$. (iv) Yes, it is.

Example 9.8. Let three elements $a = \sigma_1\sigma_2, b = 2 + \sigma_1 + \sigma_2,$ and $c = -\sigma_1\sigma_2$. (i) Determine $a(b + c)$ and $(b + c)a$. (ii) Determine $a \wedge (b + c)$. (iii) Determine $a \wedge b$ and $a \wedge c$. (iv) Is the distributivity over the exterior product fulfilled?

Solution 9.8. (i) $a(b + c) = 1 + \sigma_1 - \sigma_2 + 2\sigma_1\sigma_2$. $(b + c)a = 1 - \sigma_1 + \sigma2 + 2\sigma_1\sigma_2$. (ii) $a \wedge (b + c) = \sigma_1 - \sigma_2$. (iii) $a \wedge b = -\sigma_1 + \sigma_2$ and $a \wedge c = 0$, then $a \wedge b + a \wedge c = \sigma_1 - \sigma_2$. (iv) Yes, it is.

9.1.5. Multiplicative Inverse: a^{-1}

Definition 9.6. For an element $a \in \mathbb{G}_2$, we define $a^{-1} = \dfrac{a}{a \cdot a}$

Then

$$aa^{-1} = 1$$

Proof.

$$
\begin{aligned}
aa^{-1} &= a\,\frac{a}{a \cdot a} \\
&= \frac{a \cdot a + a \wedge a}{a \cdot a} \\
&= \frac{a \cdot a}{a \cdot a} \\
&= 1
\end{aligned}
\tag{9.4}
$$

Example 9.9. Let an element $a = \sigma_1 + 2\sigma_2 \in \mathbb{G}_2$. (i) Obtain the element aa^{-1}. (ii) Determine the geometric product aa^{-1}, the exterior product $a \wedge a^{-1}$, and the inner product $a \cdot a^{-1}$.

Solution 9.9. (i) $a^{-1} = \dfrac{a}{a \cdot a} = \dfrac{a}{5} = (\tfrac{1}{5}, \tfrac{2}{5}) \Rightarrow a^{-1} = \tfrac{1}{5}\sigma_1 + \tfrac{2}{5}\sigma_2$. (ii) $aa^{-1} = (\sigma_1 + 2\sigma_2)(\tfrac{1}{5}\sigma_1 + \tfrac{2}{5}\sigma_2) = \tfrac{1}{5}\sigma_1\sigma_1 + \tfrac{4}{5}\sigma_2\sigma_2 + \tfrac{2}{5}\sigma_1\sigma_2 - \tfrac{2}{5}\sigma_1\sigma_2$. Then $aa^{-1} = 1$, $a \cdot a^{-1} = 1$ and $a \wedge a^{-1} = 0$.

Example 9.10. Let an element $a = 2\sigma_1\sigma_2 \in \mathbb{G}_2$. Obtain the element aa^{-1}.

Solution 9.10. $a^{-1} = \dfrac{a}{a \cdot a} = -\dfrac{a}{4} = -\dfrac{\sigma_1\sigma_2}{2} \Rightarrow aa^{-1} = (-2\sigma_1\sigma_2)\dfrac{\sigma_1\sigma_2}{2} = \dfrac{2}{2} = 1$.

Example 9.11. Let an element $a = 1 - \sigma_1 + 2\sigma_1\sigma_2 \in \mathbb{G}_2$. Obtain the elements a^{-1} and aa^{-1}

Solution 9.11. $a^{-1} = \dfrac{a}{a \cdot a} = \dfrac{1 - \sigma_1 + 2\sigma_1\sigma_2}{-2 - 2\sigma_1 + 4\sigma_1\sigma_2} \Rightarrow aa^{-1} = \dfrac{(1 - \sigma_1 + 2\sigma_1\sigma_2)^2}{-2 - 2\sigma_1 + 4\sigma_1\sigma_2} = 1$.

9.1.6. Norm: $aa^{\dagger}$

Definition 9.7. For an element $a \in \mathbb{G}_2$, we define the norm of the element a as

$aa^{\dagger}$.

Note 9.3. The element $a^\dagger = (a_1 a_2 \cdots a_r)^\dagger = a_r \cdots a_2 a_1$.

Proof.

$$
\begin{aligned}
a^2 &= aa \\
&= a \cdot a + a \wedge a \\
&= a \cdot a
\end{aligned}
\tag{9.5}
$$

Example 9.12. Let an element $a = \sigma_1 \sigma_2, \in \mathbb{G}_2$. Obtain the norm of element a.

Solution 9.12. (i) From the Eq. 9.5, the norm of a is $aa^\dagger = (\sigma_1 \sigma_2)(\sigma_2 \sigma_1) = 1$.

Example 9.13. Let an element $a = 1 - \sigma_1 \sigma_2, \in \mathbb{G}_2$. Obtain the norm of element a.
Solution 9.13. (i) From the Eq. 9.5, the norm of a is $aa^\dagger = (1 - \sigma_1 \sigma_2)(1 - \sigma_2 \sigma_1) = 2$.

9.1.7. Associativity: $a(bc) = (ab)c$

Definition 9.8. For three elements a, b, and $c \in \mathbb{G}_2$, then

$$
(ab)c = a(bc)
$$

Proof.

$$
\begin{aligned}
(ab)c &= (a \cdot b + a \wedge b)c \\
&= (a \cdot b)c + (a \wedge b)c \\
&= a \cdot b \cdot c + (a \cdot b) \wedge c + (a \wedge b) \cdot c + a \wedge b \wedge c
\end{aligned}
$$

$$
\tag{9.6}
$$

$$
\begin{aligned}
a(bc) &= a(b \cdot c + b \wedge c) \\
&= a(b \cdot c) + a(b \wedge c) \\
&= a \cdot b \cdot c + a \wedge (b \cdot c) + a \cdot (b \wedge c) + a \wedge b \wedge c
\end{aligned}
$$

Note 9.4. $(a \cdot b) \wedge c + (a \wedge b) \cdot c = abc - cba$ and $a \wedge (b \cdot c) + a \cdot (b \wedge c) = abc - cba$

Example 9.14. Consider three elements $i, j,$ and $\alpha j,$ where $\alpha \in \mathbb{R}$. (i) Determine $i(j\alpha j)$. (ii) Determine $(ij)\alpha j$. (iii) Is the associativity fulfilled?

Solution 9.14. (i) $i(j\alpha j) = \sigma_1(\sigma_2 \alpha \sigma_2) = \alpha \sigma_1$. (ii) $(ij)\alpha j = (\sigma_1 \sigma_2)\alpha \sigma_2 = \alpha \sigma_1$. (iii) From the results (i) and (ii) yes, it is.

9.1.8. Rigid Body Motions

Given a **vector** $a = a_1\sigma_1 + a_2\sigma_2 \in \mathbb{G}_2$, we define a **rotation** or **reflection** of vector a as Ia or IIa, where $I = \sigma_1 \wedge \sigma_2$.

Example 9.15. Let an element $a = \sigma_1 + \sigma_2 \in \mathbb{G}_2$. (i) Determine Ia. (ii) Determine aI. (iii) Explain (i) and (ii).

Solution 9.15. (i) $Ia = \sigma_1\sigma_2 a = -\sigma_2 + \sigma_1 = \sigma_1 - \sigma_2$. (ii) $aI = a\sigma_1\sigma_2 = \sigma_2 - \sigma_1$. (iii) From these results, (i) is a **rotation** of $\frac{\pi}{2}$ in the **clockwise** direction and (ii) is a **rotation** of $\frac{\pi}{2}$ in the **counter-clockwise** direction.

Example 9.16. Let an element $a = \sigma_1 + \sigma_2 \in \mathbb{G}_2$. (i) Determine IIa. (ii) Determine aII. (iii) Explain (i) and (ii). (i) $IIa = \sigma_1\sigma_2\sigma_1\sigma_2 a = -\sigma_1 - \sigma_2$. (ii) $aII = a\sigma_1\sigma_2\sigma_1\sigma_2 = -\sigma_1 - \sigma_2$. (iii) From these results, (i) is a **reflection** of π in the **clockwise** direction and (ii) is a **reflection** of π in the **counter-clockwise** direction.

9.2. GEOMETRIC ALGEBRA $\mathbb{G}_3$ in $\mathbb{R}^3$

Definition 9.9. We define the unitary associative algebra named **Geometric Algebra** [54] $\mathbb{G}_3 = \mathbb{G}_3(\mathbb{R}^3)$, and eight elements: scalars, $\sigma_1, \sigma_2, \sigma_3$ vectors, $\sigma_1 \wedge \sigma_2$, $\sigma_1 \wedge \sigma_3$, $\sigma_2 \wedge \sigma_3$ **bivectors**, and $\sigma_1 \wedge \sigma_2 \wedge \sigma_3$ **trivectors** (or **equivalently** $\sigma_1\sigma_2\sigma_3$), which form an **orthonormal** basis $(1, \sigma_1, \sigma_2, \sigma_1 \wedge \sigma_2, \sigma_1 \wedge \sigma_3, \sigma_2 \wedge \sigma_3, \sigma_1 \wedge \sigma_2 \wedge \sigma_3)$, that meets $\sigma_i \wedge \sigma_i = 1$ and $\sigma_i \wedge \sigma_j = -\sigma_j \wedge \sigma_i$. For the algebra $\mathbb{G}_3$ there is an arbitrary element

$$v = v_0 + v_1\sigma_1 + v_2\sigma_2 + v_{12}\sigma_1 \wedge \sigma_2 + v_{13}\sigma_1 \wedge \sigma_3 + v_{23}\sigma_2 \wedge \sigma_3 + v_{123}\sigma_1 \wedge \sigma_2 \wedge \sigma_3 \in \mathbb{G}_3.$$

From two elements $a = a_0 + a_1\sigma_1 + a_2\sigma_2 + a_3\sigma_3 + a_4\sigma_1\sigma_2, +a_5\sigma_1\sigma_3 + a_6\sigma_2\sigma_3 + a_7\sigma_1\sigma_2\sigma_3$ and $b = b_0 + b_1\sigma_1 + b_2\sigma_2 + b_3\sigma_3 + b_4\sigma_1\sigma_2, +b_5\sigma_1\sigma_3 + b_6\sigma_2\sigma_3 + b_7\sigma_1\sigma_2\sigma_3 \in \mathbb{G}_2$, the **geometric product** (Eq. (1)) is defined as

$$ab = (a_0 + a_1\sigma_1 + a_2\sigma_2 + a_3\sigma_3 + +a_4\sigma_1\sigma_2, +a_5\sigma_1\sigma_3 + a_6\sigma_2\sigma_3 + a_7\sigma_1\sigma_2\sigma_3) \\ (b_0 + b_1\sigma_1 + b_2\sigma_2 + b_3\sigma_3 + b_4\sigma_1\sigma_2, +b_5\sigma_1\sigma_3 + b_6\sigma_2\sigma_3 + b_7\sigma_1\sigma_2\sigma_3) \quad (1)$$

Alternatively, the geometric product is defined by $ab = a \cdot b + a \wedge b$, where the term $a \cdot b$ is called **inner product** (Def. 1.2) and the term $a \wedge b$ is called **exterior product** (Def. 9.10).

Note 9.5. The exterior product $\sigma_1\sigma_3$ is only used with the elements σ_i of the orthogonal base, this expression should **never** be used with the elements of $\mathbb{G}_2$, *i.e.* $a \wedge b$.

Note 9.6. The geometrical description of the **trivector** $\sigma_1 \wedge \sigma_2 \wedge \sigma_3$ is the **oriented** volume spanned by the vectors σ_1, σ_2 and σ_3 (Fig. **9.2**).

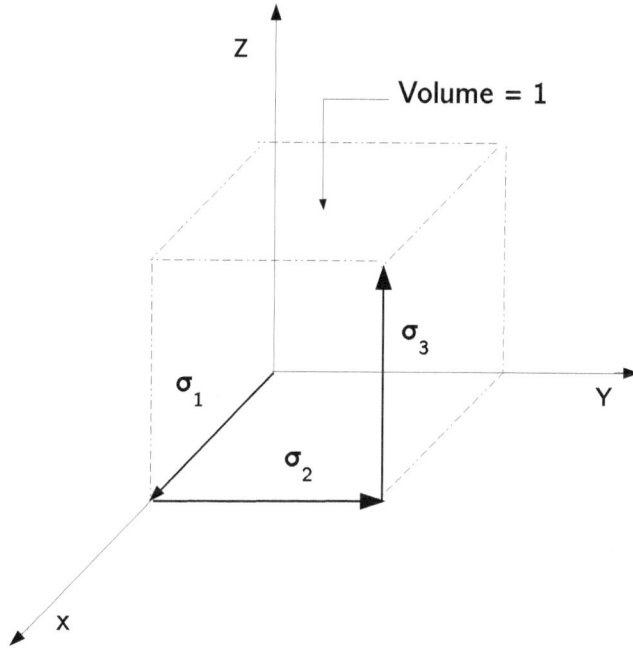

Fig. (9.2). The oriented volume formed by the vectors σ_1, σ_2, and σ_3.

Example 9.17. Let two elements $a = \sigma_1 - 2\sigma_2 + 3\sigma_3 - \sigma_1\sigma_2$ and $b = 2\sigma_1 + 3\sigma_2 + \sigma_3 \in \mathbb{G}_3$. (i) Obtain the geometric product ab and ba. (ii) Obtain $a \wedge b = \frac{1}{2}(ab - ba)$. (iii) Obtain $a \cdot b = \frac{1}{2}(ab + ba)$. (iv) Explain the results obtained in (iii).

Solution 9.17. (i) $ab = (\sigma_1 - 2\sigma_2 + 3\sigma_3 - \sigma_1\sigma_2)(2\sigma_1 + 3\sigma_2 + \sigma_3) = -1 - 3\sigma_1 + 2\sigma_2 + 7\sigma_1\sigma_2 - 5\sigma_1\sigma_3 - 11\sigma_2\sigma_3 - \sigma_1\sigma_2\sigma_3$ $ba = (2\sigma_1 + 3\sigma_2 + \sigma_3)(\sigma_1 - 2\sigma_2 + 3\sigma_3 - \sigma_1\sigma_2) = -1 + 3\sigma_1 - 2\sigma_2 - 7\sigma_1\sigma_2 + 5\sigma_1\sigma_3 + 11\sigma_2\sigma_3 - \sigma_1\sigma_2\sigma_3$.
(ii) $a \wedge b = -6\sigma_1 + 4\sigma_2 + 14\sigma_1\sigma_2 - 10\sigma_1\sigma_3 - 22\sigma_2\sigma_3$. (iii) $a \cdot b = -1 - \sigma_1\sigma_2\sigma_3$. (iv) The **geometric product** is the addition of the **inner product** and the **exterior product**, but the **inner product** is **not** the real part of the geometric product.

9.2.1. Exterior Product: $a \wedge b$

Definition 9.10. For two elements a and $b \in \mathbb{G}_3$, we define (Def. 9.2.1)

$$a \wedge b = \frac{1}{2}(ab - ba).$$

Example 9.18. Let two elements $a = \sigma_1\sigma_2$ and $b = \sigma3 \in \mathbb{G}_3$. Obtain the exterior product $a \wedge b = \frac{1}{2}(ab - ba)$.

Solution 9.18. $ab = \sigma_1\sigma_2\sigma_3$, $ba = \sigma_1\sigma_2\sigma_3$, then $a \wedge b = 0$.

9.2.2. Inner Product: $a \cdot b$

Definition 9.11. For two elements a and $b \in \mathbb{G}_3$, we define

$$a \cdot b = \frac{1}{2}(ab + ba).$$

Example 9.19. Consider two elements $a = \sigma_1\sigma_2\sigma_3, b = \sigma_1 - \sigma_2\sigma_1\sigma_3 \in \mathbb{G}_3$. (i) Obtain the geometric products ab and ba. (ii) Determine $a \cdot b$. (iii) Determine $a \wedge b$.

Solution 9.19. (i) $ab = (\sigma_1\sigma_2\sigma_3)(\sigma_1 - \sigma_2\sigma_1\sigma_3) = \sigma_2\sigma_3 - 1,\ ba = \sigma_2\sigma_3 - 1.$ (ii) $a \cdot b = \sigma_2\sigma_3 - 1$. (iii) $a \wedge b = 0$.

Example 9.20. Consider two elements $a = \sigma_1\sigma_2$, and $b = \sigma_3 \in \mathbb{G}_3$. (i) Obtain the geometric products ab and ba. (ii) From Def. 9.1.1, determine $a \cdot b$. (iii) From Def. 9.1.0, determine $a \wedge b$. (i) $ab = \sigma_1\sigma_2\sigma_3$. $ba = \sigma_1\sigma_2\sigma_3$. (ii) $a \cdot b = \sigma_1\sigma_2\sigma_3$. (iii) $a \wedge b = 0$.

Example 9.21. Consider two elements $a = 1 + \sigma_1 + \sigma_2 - \sigma_2\sigma_3$, and $b = \sigma_1\sigma_2 \in \mathbb{G}_3$. (i) Obtain the geometric products ab and ba. (ii) From Def. 2.2, determine $a \cdot b$. (iii) From Def. 2.1, determine $a \wedge b$.

Solution 9.21. (i) $ab = (1 + \sigma_1 + \sigma_2 - \sigma_2\sigma_3)(\sigma_1\sigma_2) = \sigma_1 - \sigma_2 + \sigma_1\sigma_2 - \sigma_1\sigma_3.$ $ba = (\sigma_1\sigma_2)(1 + \sigma_1 + \sigma_2 - \sigma_2\sigma_3) = -\sigma_1 + \sigma_2 + \sigma_1\sigma_2 - \sigma_1\sigma_3.$ (ii) $a \cdot b = \sigma_1\sigma_2 - \sigma_1\sigma_3$. (iii) $a \wedge b = \sigma_1 - \sigma_2$.

9.2.3. Distributivity: $a(b + c)$

Definition 9.12. For three elements a, b, and $c \in \mathbb{G}_3$, then

$$a(b+c) = ab + ac$$

Proof.

$$
\begin{aligned}
ab + ac &= a \cdot b + a \wedge b + a \cdot c + a \wedge c \\
&= a \cdot b + a \cdot c + a \wedge b + a \wedge c \\
&= a \cdot (b+c) + a \wedge (b+c) \\
&= a(b+c)
\end{aligned}
\tag{9.8}
$$

Example 9.22. Consider three elements $a = 1 + \sigma_1, b = \sigma_1\sigma_3$, and $c = \sigma_1 + \sigma_2$. (i) Determine $a(b+c)$. (ii) Determine $ab + ac$. (iii) Is the distributivity fulfilled?

Solution 9.22. (i) $a(b+c) = (1+\sigma_1)(\sigma_1 + \sigma_2 + \sigma_1\sigma_2) = 1 + \sigma_1 + \sigma_2 + \sigma_3 + \sigma_1\sigma_2 + \sigma_1\sigma_3$. (ii) $ab + ac = (1+\sigma_1)(\sigma_1\sigma_3) + (1+\sigma_1)(\sigma_1 + \sigma_2 = 1 + \sigma_1 + \sigma_2 + \sigma_3 + \sigma_1\sigma_2 + \sigma_1\sigma_3$. (iii) From (i) and (ii) yes, it is.

9.2.4. Distributivity: $a \wedge (b+c)$

Definition 9.13. For three elements a, b, and $c \in \mathbb{G}_3$, then

$$a \wedge (b+c) = a \wedge b + a \wedge c$$

Proof.

$$
\begin{aligned}
a \wedge (b+c) &= -ab - ac + a \cdot (b+c) \\
&= -ab - ac + a \cdot b + a \cdot c \\
&= (-ab + a \cdot b) + (-ac + a \cdot c) \\
&= a \wedge b + a \wedge c
\end{aligned}
\tag{9.9}
$$

Example 9.23. Consider three elements $a = \sigma_1\sigma_2\sigma_3, b = \sigma_3\sigma_2$, and $c = \sigma_1\sigma_2$. (i) Determine $a(b+c)$ and $(b+c)a$. (ii) Determine $a \wedge (b+c)$. (iii) Determine $a \wedge b$ and $a \wedge c$. (iv) Is the distributivity over the exterior product fulfilled?

Solution 9.23. (i) $a(b+c) = (\sigma_1\sigma_2\sigma_3)(\sigma_1\sigma_2 + \sigma_3\sigma_2) = -\sigma_3 + \sigma_1$. $(b+c)a = (\sigma_1\sigma_2 + \sigma_3\sigma_2)(\sigma_1\sigma_2\sigma_3) = -\sigma_3 + \sigma_1$. (ii) $a \wedge (b+c) = 0$. (iii) $a \wedge b = 0$, and $a \wedge c = 0$, then $a \wedge b + a \wedge c = 0$. (iv) Yes, it is.

9.2.5. Multiplicative Inverse: a^{-1}

Definition 9.14. For an element $a \in \mathbb{G}_3$, we define $a^{-1} = \dfrac{a}{a \cdot a}$

Then

$$aa^{-1} = 1$$

Proof.

$$
\begin{aligned}
aa^{-1} &= a\frac{a}{a \cdot a} \\
&= \frac{a \cdot a + a \wedge a}{a \cdot a} \\
&= \frac{a \cdot a}{a \cdot a} \\
&= 1
\end{aligned}
\tag{9.10}
$$

Example 9.24. Let an element $a = \sigma_1 + 2\sigma_2\sigma_3 \in \mathbb{G}_3$. (i) Obtain the element aa^{-1}. (ii) Determine the geometric product aa^{-1}, the exterior product $a \wedge a^{-1}$, and the inner product $a \cdot a^{-1}$.

Solution 9.24. (i) $a^{-1} = \frac{\sigma_1 + 2\sigma_2\sigma_3}{-3 + 4\sigma_1\sigma_2\sigma_3}$. (ii) $aa^{-1} = \frac{(\sigma_1 + 2\sigma_2\sigma_3)^2}{-3 + 4\sigma_1\sigma_2\sigma_3}$. Then $aa^{-1} = 1$, $a \cdot a^{-1} = 1$ and $a \wedge a^{-1} = 0$.

Example 9.25. Let an element $a = \sigma_1\sigma_2\sigma_3 \in \mathbb{G}_3$. Obtain the element aa^{-1}.

Solution 9.25. $a^{-1} = \frac{a}{a \cdot a} = -a = -\sigma_1\sigma_2\sigma_3 \Rightarrow aa^{-1} = (\sigma_1\sigma_2\sigma_3)(-\sigma_1\sigma_2\sigma_3) = 1$.

9.2.6. Norm: $aa^\dagger$

Definition 9.15. For an element $a \in \mathbb{G}_3$, we define the norm of an element a as

$$aa^\dagger.$$

Note 9.7. The element $a^\dagger = (a_1 a_2 \cdots a_r)^\dagger = a_r \cdots a_2 a_1$.

Proof.

$$
\begin{aligned}
a^2 &= aa \\
&= a \cdot a + a \wedge a \\
&= a \cdot a
\end{aligned}
\tag{9.11}
$$

Example 9.26. Consider an element $a = \sigma_1\sigma_2\sigma_3, \in \mathbb{G}_3$. Obtain the norm of the element a.

Solution 9.26. From the Eq. 9.11, the norm of a is $aa^\dagger = (\sigma_1\sigma_2\sigma_3)(\sigma_3\sigma_2\sigma_1) = 1$.

Example 9.27. Consider an element $a = 2 - \sigma_1\sigma2 + \sigma_1\sigma_2\sigma_3, \in \mathbb{G}_3$. Obtain the norm of the element a.

Solution 9.27. From the Eq. 9.11, the norm of a is $aa^\dagger = (2 - \sigma_1\sigma2 + \sigma_1\sigma_2\sigma_3)$ $(2 - \sigma_2\sigma1 + \sigma_3\sigma_2\sigma_1) = 4 + 2\sigma_1\sigma_2 + 2\sigma_3\sigma_2\sigma_1 - 2\sigma_1\sigma_2 + 1 - \sigma_3 + 2\sigma_1\sigma_2\sigma_3 +$ $\sigma_3 + 1 = 4$.

9.2.7. Associativity: $a(bc) = (ab)c$

Definition 9.16. For three elements a, b, and $c \in \mathbb{G}_3$, then

$$(ab)c = a(bc)$$

Proof.

$$
\begin{aligned}
(ab)c &= (a \cdot b + a \wedge b)c \\
&= (a \cdot b)c + (a \wedge b)c \\
&= a \cdot b \cdot c + (a \cdot b) \wedge c + (a \wedge b) \cdot c + a \wedge b \wedge c
\end{aligned}
\tag{9.12}
$$

$$
\begin{aligned}
a(bc) &= a(b \cdot c + b \wedge c) \\
&= a(b \cdot c) + a(b \wedge c) \\
&= a \cdot b \cdot c + a \wedge (b \cdot c) + a \cdot (b \wedge c) + a \wedge b \wedge c
\end{aligned}
$$

Note 9.8. $(a \cdot b) \wedge c + (a \wedge b) \cdot c = abc - cba$ and $a \wedge (b \cdot c) + a \cdot (b \wedge c) = abc - cba$

Example 9.28. Consider three elements $i, j,$ and αj, where $\alpha \in \mathbb{R}$. (i) Determine $i(j\alpha j)$. (ii) Determine $(ij)\alpha j$. (iii) Is the associativity fulfilled?

Solution 9.28. (i) $i(j\alpha j) = \sigma_1(\sigma_2\alpha\sigma_2) = \alpha\sigma_1$. (ii) $(ij)\alpha j = (\sigma_1\sigma_2)\alpha\sigma_2 = \alpha\sigma_1$. (iii) From these results yes, it is.

9.3. CASE STUDY: EQUATION OF A LINE ON THE PLANE

Example 9.29. Given a **vector** v and a point x_0 in the plane, what is the equation of the line passing through the point x_0 in the direction of the vector v (Fig.**9.3**)? (Case adapted with permission of the author [17]).

The line $L_{x_0}(v)$ is given by

$$L_{x_0}(v) \quad := \{x \mid (x - x_0) \wedge v = 0\} \tag{9.13}$$

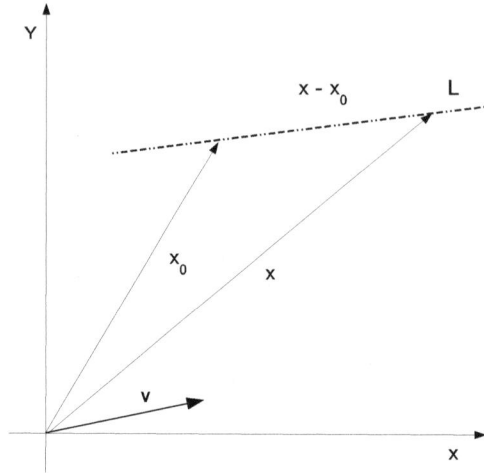

Fig. (9.3). The line L_{x_0} through the point x_0 in the direction v.

Example 9.30. Given a vector v and a point x_0 in the plane, what is the equation of the line passing through the point $x_0 = (1,2)$ in the direction of the vector $v = (1,1)$?

The line $L_{x_0}(v)$ is given by

$$L_{x_0=(1,2)}(v) \quad := \{x \mid (x - x_0) \wedge v = 0\} \tag{9.14}$$

$$\begin{aligned} [(x_1\sigma_1 + x_2\sigma_2) - (\sigma_1 + 2\sigma_2)] \wedge (\sigma_1 + \sigma_2) \quad &= 0 \\ \wedge (\sigma_1 + \sigma_2) \quad &= 0 \end{aligned} \tag{9.15}$$

The exterior product $(x - x_0) \wedge v = \frac{1}{2}[(x - x_0)v - v(x - x_0)]$,

$$[(x_1 - 1)\sigma_1 + (x_2 - 2)\sigma_2](\sigma_1 + \sigma_2) \quad = (x_1 + x_2 - 3) + (x_1 - x_2 + 1)\sigma_1\sigma_2 \tag{2}$$

$$(\sigma_1 + \sigma_2)[(x_1 - 1)\sigma_1 + (x_2 - 2)\sigma_2] \quad = (x_1 + x_2 - 3) + (x_2 - x_1 - 1)\sigma_1\sigma_2$$

From Eqs. 9.16, $(x - x_0) \wedge v = x_1 - x_2 - 1 \Rightarrow x_1 = x_2 - 1$. So, the points with the form $(x_2 - 1, x_2)$ are the solution. Note that the points $(1,2)$ and $(0,1)$ meet the line $L_{x_0}(v)$.

9.4. CASE STUDY: EQUATION OF A PLANE IN SPACE

Example 9.31. Given two **vectors** v and u, and a point x_0 in space, what is the equation of the plane passing through the point x_0 over the plane generated by the vectors v and u? (Fig. **9.4**) (Case adapted with permission of the author [17]).

The plane $P_{x_0}(u \wedge v)$ is given by

$$P_{x_0}(u \wedge v) \quad := \{x \mid (x - x_0) \wedge (u \wedge v) = 0\} \tag{9.17}$$

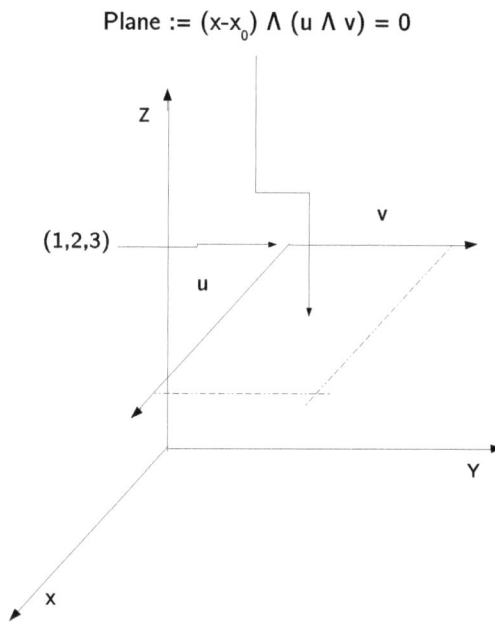

Fig. (9.4). The plane P_{x_0} through the point x_0 in the direction of the bivector $u \wedge v$.

Example 9.32. Given two vectors v, and u, and a point x_0 in the plane, what is the equation of the plane passing through the point $x_0 = (1,2,3)$ in the plane generated by the vectors $v = (1,0,0)$ and $u = (0,1,0)$? The plane $P_{x_0}(u,v)$ is given by

$$P_{x_0=(1,2,3)}(u \wedge v) \quad := \{x \mid (x - x_0) \wedge (u \wedge v) = 0\} \tag{9.18}$$

$$[(x_1\sigma_1 + x_2\sigma_2 + x_3\sigma_3) - (\sigma_1 + 2\sigma_2 + 3\sigma_3)] \wedge (\sigma_1\sigma_2) = 0$$
$$[(x_1 - 1)\sigma_1 + (x_2 - 2)\sigma_2 + (x_3 - 3)\sigma_3] \wedge (\sigma_1\sigma_2) = 0$$

(9.19)

From Eq. 9.19, $[(x_1 - 1)\sigma_1 + (x_2 - 2)\sigma_2 + (x_3 - 3)\sigma_3][\sigma_1\sigma_2] - [\sigma_1\sigma_2][(x_1 - 1)\sigma_1 + (x_2 - 2)\sigma_2 + (x_3 - 3)\sigma_3] = 0$. So, the points with the form $(2 - x_2, 1 - x_1, 3)$ are the solution. Note that the point (1,2,3) meets the plane $P_{x_0=(1,2,3)}(u \wedge v)$.

9.5. EXERCISES

Exercise 9.1. Let $w = 3dx + 2dy$, and $\eta = e^x dx + 2dy$ [55]. (i) Compute the exterior product $w \wedge \eta$. (ii) Compute $d(w \wedge \eta)$.

Exercise 9.2. Let $w = (x_1 + x_3^2)dx_1 \wedge dx_2$ [56]. Compute dw.

Exercise 9.3. Compute $dx \wedge dy = (-r\sin\theta d\theta + \cos\theta dr) \wedge (r\cos\theta d\theta + \sin\theta dr)$ [57].

Exercise 9.4. Let $w = xdx + yzdy + x^2ydz$ and $\eta = xydz$ [58]. (i) Compute dw. (ii) Compute $d(w)$. (iii) Compute $w \wedge \eta$. (iv) Compute $d(w \wedge \eta)$. (v) Explain (iv).

CHAPTER 10

Theorems of Differential Forms

Abstract: This chapter reviews the main aspects of the Differential forms and their application to facilitate the resolution of the integrals involved in Green's theorem, Stokes' theorem, and Gauss' theorem.

Keywords: 0-form, 1-form, 2-form, 3-form, p-form, Exterior product, Exterior derivative, Geometric algebra, Grassmann algebra, Gauss' theorem, Green's theorem, Differential forms, Differentiation, Line integral, Oriented curve, Stokes' theorem, Surface integral.

10.1. DIFFERENTIAL FORMS

The differential forms are defined over the Geometric product (Sect. 9.1).

Definition 10.1. Given an open set $\Omega \in \mathbb{R}^n$, the following differential k-forms can be defined [59]:

1. A 0-form on an open subset Ω of R^n, where the real-valued functions f_i are defined $f_i: \Omega \rightarrow \mathbb{R}$, is expressed by

$$\omega = \sum_{i=0}^{n} f_i(x_1, x_2, \cdots, x_n).$$

For example the real-value function $f(x, y) = 6y^3 + xy$ is a 0-form.

2. A 1-form on an open subset Ω of R^n, where the real-valued functions f_i are defined $f_i: \Omega \rightarrow \mathbb{R}$, is expressed by

$$\omega = \sum_{i=0}^{n} f_i(x_1, x_2, \cdots, x_n) dx_i.$$

For example the function ω defined as $\omega = xy^3 dx + yzdy - zsinyzdz$ is a 1-form.

3. A 2-form on an open subset Ω of R^n, where the real-valued functions f_{ij} are defined $f_{ij} : \Omega \to \mathbb{R}$, is expressed

$$\omega = \Sigma_{i,j=0}^{n} f_{ij}(x_1, x_2, \cdots, x_n) dx_i dx_j.$$

For example the function ω defined as $\omega = xy dx dy + yz dy dz - z\cos yz dz dx$ is a 2-form.

4. A 3-form on an open subset Ω of R^n, where the real-valued functions f_{ijk} are defined $f_{ijk} : \Omega \to \mathbb{R}$, is expressed

$$\omega = \Sigma_{i,j,k=0}^{n} f_{ijk}(x_1, x_2, \cdots, x_n) dx_i dx_j dx_k.$$

For example the function ω defined as $\omega = xyz dx dy dz + yz\cos x dy dz dx - z\cos yz dy dz dx$ is a 3-form.

5. A p-form on an open subset Ω of R^n, where the real-valued functions $f_{ijk \cdots p}$ are defined $f_{ijk \cdots p} : \Omega \to \mathbb{R}$, is expressed

$$\omega = \Sigma_{i,j,k,\cdots,p=0}^{n} f_{ijkp}(x_1, x_2, \cdots, x_n) dx_i dx_j dx_k \cdots dx_p.$$

For example the function ω defined as $\omega = \sin x dx dy dz dp$ is a p-form.

10.1.1. Algebra of Differential Forms

Definition 10.2. If the function w is a k-form and η is a l-form on K, with $0 \leq k + l \leq 3$ [25], then the functions meet the **exterior product** properties (Sects 9.1, 9.2. In short notation: $dx \wedge dy = dxdy$, $dx \wedge dy = dxdy = -dy \wedge dx = -dydx$ and $dx \wedge dx = dxdx = 0$.

Example 10.1. Let $w = P(x, y, z)dx + Q(x, y, z)dy$ on $\mathbb{R}^3$ be a 1-form. Determine its exterior derivative dw.

Solution 10.1.

$$
\begin{aligned}
dw &= \\
&= d[Pdx + Qdy] \\
&= d(P \wedge dx) + d(Q \wedge dy) \\
&= \left(\frac{\partial P}{\partial x}dx + \frac{\partial P}{\partial y}dy + \frac{\partial P}{\partial z}dz\right) \wedge dx + \left(\frac{\partial Q}{\partial x}dx + \frac{\partial Q}{\partial y}dy + \frac{\partial Q}{\partial z}dz\right) \wedge dy \\
&= \left(\frac{\partial P}{\partial x}dxdx\right) + \left(\frac{\partial P}{\partial y}dydx\right) + \left(\frac{\partial P}{\partial z}dzdx\right) + \left(\frac{\partial Q}{\partial x}dxdy\right) + \left(\frac{\partial Q}{\partial y}dydy\right) + \left(\frac{\partial Q}{\partial z}dzdy\right) \\
&= \left(\frac{\partial P}{\partial x}dxdx\right) - \left(\frac{\partial P}{\partial y}dxdy\right) + \left(\frac{\partial P}{\partial z}dzdx\right) + \left(\frac{\partial Q}{\partial x}dxdy\right) - \left(\frac{\partial Q}{\partial y}dxdy\right) + \left(\frac{\partial Q}{\partial z}dzdy\right) \\
&= -\left(\frac{\partial P}{\partial y}dxdy\right) + \left(\frac{\partial P}{\partial z}dzdx\right) + \left(\frac{\partial Q}{\partial x}dxdy\right) - \left(\frac{\partial Q}{\partial z}dydz\right) \\
&= \left(\frac{\partial Q}{\partial x} - \frac{\partial P}{\partial y}\right)dxdy + \frac{\partial P}{\partial z}dzdx - \frac{\partial Q}{\partial z}dydz
\end{aligned}
\tag{10.1}
$$

Example 10.2. Let $w = xydx + e^z dy + xdz$ [60]. (i) Determine dw. (ii) From this result what can you say about w and dw?

Solution 10.2. (i)

$$
\begin{aligned}
dw &= \\
&= \frac{\partial xy}{\partial x}dx + \frac{\partial xy}{\partial y}dx + \frac{\partial xy}{\partial z}dx + \frac{\partial e^z}{\partial x}dy + \frac{\partial e^z}{\partial y}dy + \frac{\partial e^z}{\partial z}dy + \frac{\partial x}{\partial x}dz + \frac{\partial x}{\partial y}dz + \frac{\partial x}{\partial z}dz \\
&= ydxdx + xdydx + 0dzdx + 0dxdy + 0dydy + e^z dzdy + dxdz + 0dydz + 0dzdz \\
&= xdydx + e^z dzdy + dxdz \\
&= -xdxdy - e^z dydz + dxdz.
\end{aligned}
\tag{10.2}
$$

(ii) The derivative of a k-form is a $(k + 1)$-form if $k < 3$.

Example 10.3. Let $w = xdx - ydy$ and $\eta = xdydz + zdxdy$ [61]. Determine $w \wedge \eta$.

Solution 10.3.

$$
\begin{aligned}
w \wedge \eta &= \\
&= (xdx - ydy) \wedge (xdy \wedge dz + zdx \wedge dy) \\
&= [(xdx - ydy) \wedge (xdy \wedge dz)] + [(xdx - ydy) \wedge (zdx \wedge dy)] \\
&= (x^2 dx \wedge dy \wedge dz) - (xydy \wedge dy \wedge dz) + (xzdx \wedge dx \wedge dy) - (yzdy \wedge dx \wedge dy) \\
&= [x^2 dx \wedge dy \wedge dz] - [xy(dy \wedge dy) \wedge dz] + [xz(dx \wedge dx) \wedge dy] - [yz(dy \wedge dx) \wedge dy] \\
&= x^2 dxdydz.
\end{aligned}
\tag{10.3}
$$

In short notation

$$
\begin{aligned}
w \wedge \eta \;=\; & \\
=\; & (xdx - ydy)(xdydz + zdxdy) \\
=\; & [(xdx - ydy)(xdydz)] + [(xdx - ydy)(zdxdy] \\
=\; & (x^2 dxdydz) - (xydydydz) + (xzdxdxdy) - (yzdydxdy) \\
=\; & [x^2 dxdydz] - [xy(dydy)dz] + [xz(dxdx)dy] - [yz(dydx)dy] \\
=\; & x^2 dxdydz.
\end{aligned} \tag{10.4}
$$

Example 10.4. Let $\eta = P(x,y,z)dydz + Q(x,y,z)dzdx + R(x,y,z)dx\,dy$ on $\mathbb{R}^3$ be a 1-form [61]. Determine its exterior derivative $d\eta$.

Solution 10.4.

$$
\begin{aligned}
d\eta \;=\; & d(Pdydz) + d(Qdzdx) + d(Rdxdy) \\
=\; & d(P \wedge dydz) + d(Q \wedge dzdx) + d(R \wedge dxdy) \\
=\; & (\tfrac{\partial P}{\partial x}dx + \tfrac{\partial P}{\partial y}dy + \tfrac{\partial P}{\partial z}dz) \wedge (dy \wedge dz) + (\tfrac{\partial Q}{\partial x}dx + \tfrac{\partial Q}{\partial y}dy + \tfrac{\partial Q}{\partial z}dz) \wedge (dz \wedge dx) \\
& +(\tfrac{\partial R}{\partial x}dx + \tfrac{\partial R}{\partial y}dy + \tfrac{\partial R}{\partial z}dz) \wedge (dx \wedge dy) \\
=\; & \tfrac{\partial P}{\partial x}dx(dy \wedge dz) + \tfrac{\partial P}{\partial y}dy \wedge (dy \wedge dz) + \tfrac{\partial P}{\partial z}dz(dy \wedge dz) + \tfrac{\partial Q}{\partial x}dx(dz \wedge dx) \\
& + \tfrac{\partial Q}{\partial y}dy(dz \wedge dx) + \tfrac{\partial Q}{\partial z}dz(dz \wedge dx) + \tfrac{\partial R}{\partial x}dx(dx \wedge dy) + \tfrac{\partial R}{\partial y}dy \wedge (dx \wedge dy) \\
& + \tfrac{\partial R}{\partial z}dz(dx \wedge dy) \\
=\; & \tfrac{\partial P}{\partial x}dxdydz + \tfrac{\partial P}{\partial y}dydydz + \tfrac{\partial P}{\partial z}dzdydz + \tfrac{\partial Q}{\partial x}dxdzdx + \tfrac{\partial Q}{\partial y}dydzdx + \tfrac{\partial Q}{\partial z}dzdzdx \\
& + \tfrac{\partial R}{\partial x}dxdxdy + \tfrac{\partial R}{\partial y}dydxdy + \tfrac{\partial R}{\partial z}dzdxdy \\
=\; & (\tfrac{\partial P}{\partial z} + \tfrac{\partial Q}{\partial y} + \tfrac{\partial R}{\partial x})\,dxdydz
\end{aligned} \tag{10.5}
$$

10.2. LINE INTEGRAL ON $\mathbb{G}_3$

Definition 10.3. We define a **line integral** over an **oriented** curve T, using the map $T \colon \mathbb{R} \to \mathbb{R}^3;\ (x(t), y(t), z(t))$, the function w that is C^1 class, and a 2-form [25], $w \colon \mathbb{R}^3 \to \mathbb{R}^3;\ P(x,y,z)dx + Q(x,y,z)dy + R(x,y,z)dz$.

$$
\begin{aligned}
w(t) \;=\; & \int_W w \\
=\; & \int_W Pdx + Qdy + Rdz \\
=\; & \int_{t_0}^{t_1} P\tfrac{dx}{dt} + Q\tfrac{dy}{dt} + R\tfrac{dz}{dt}\ dt \\
=\; & \int_{t_0}^{t_1} P \circ T(t)\tfrac{dx}{dt} + Q \circ T(t)\tfrac{dy}{dt} + R \circ T(t)\tfrac{dz}{dt}\ dt
\end{aligned} \tag{10.6}
$$

Example 10.5. Let the function $w = xy^2 dx + 3y^3 dy + dz$ be on space and the oriented curve C described as $T(t) = (t, t^2, 2), t \in [0,1]$. (i) Determine $P(x, y, z)$, $Q(x, y, z)$, and $R(x, y, z)$. (ii) Determine $x(t)$, $y(t)$, and $z(t)$. (iii) Determine $P \circ T(t)$, $Q \circ T(t)$, and $R \circ T(t)$. (iv) Determine $x'(t)$, $y'(t)$, and $z'(t)$. (v) Compute the integral over the curve $T(t)$.

Solution 10.5. (i) $P(x, y, z) = xy^2$, $Q(x, y, z) = y^3$, and $R(x, y, z) = 2$. (ii) $x(t) = t, y(t) = t^2$, and $z(t) = 2$. (iii) $P \circ T(t) = t^5, Q \circ T(t) = 3t^6$, and $R \circ T(t) = 2$. (iv) $x'(t) = 1, y'(t) = 2t$, and $z'(t) = 0$. (v) $\int_{t_0}^{t_1} (P \circ T(t)) x' + (Q \circ T(t)) y' + (R \circ T(t)) z' \, dt = \int_0^1 t^5(1) + 3t^6(2t) + 2(0) \, dt = \int_0^1 t^5 + 6t^7 \, dt = \frac{1}{6} + \frac{6}{8} = \frac{44}{48}$.

10.3. SURFACE INTEGRAL ON $\mathbb{G}_3$

Definition 10.4. We define the **surface integral** over the **oriented** surface T, using the map $T: \mathbb{R}^2 \to \mathbb{R}^3$; $(x(u, v), y(u, v), z(u, v))$, a function w that is C^1 class, and a 1-form [25], $w: \mathbb{R}^3 \to \mathbb{R}^3$; $P(x, y, z)dxdy + Q(x, y, z)dydz + R(x, y, z)dzdx$.

$$
\begin{aligned}
w(T) &= \iint_S w \\
&= \iint_{u,v-region} P dy \wedge dz + Q dz \wedge dx + R dx \wedge dy \\
&= \int_{u_0}^{u_1} \int_{v_0}^{v_1} P \cdot \frac{\partial(y,z)}{\partial(u,v)} + Q \cdot \frac{\partial(z,x)}{\partial(u,v)} + R \cdot \frac{\partial(x,y)}{\partial(u,v)} \, dv \, du \\
&= \int_{u_0}^{u_1} \int_{v_0}^{v_1} P \circ T(u,v) \frac{\partial(y,z)}{\partial(u,v)} + Q \circ T(u,v) \frac{\partial(z,x)}{\partial(u,v)} + R \circ T(u,v) \frac{\partial(x,y)}{\partial(u,v)} \, dv \, du
\end{aligned}
\tag{10.7}
$$

Example 10.6. Let the function $w = x^2 dxdy$ be on space [25], the oriented surface S defined by $T(u, v) = (\sin u \cos v, \sin u \sin v, \cos u), u \in [0, \frac{\pi}{2}]$, and $v \in [0, 2\pi]$. (i) Determine $P(x, y, z)$, $Q(x, y, z)$, and $R(x, y, z)$. (ii) Determine $x(u, v)$, $y(u, v)$, and $z(u, v)$. (iii) Determine $P \circ T(u, v)$, $Q \circ T(u, v)$, and $R \circ T(u, v)$. (iv) Determine $\frac{\partial(x,y)}{\partial(u,v)}$, $\frac{\partial(y,z)}{\partial(u,v)}$, and $\frac{\partial(z,x)}{\partial(u,v)}$. (v) Compute the integral over the surface $T(u, v)$.

Solution 10.6. (i) $P(x, y, z) = 0$, $Q(x, y, z) = 0$, and $R(x, y, z) = x^2$. (ii) $x(u, v) = \sin u \cos v, y(t) = \sin u \sin v$, and $z(t) = \cos u$. (iii) $P \circ T(u, v) = 0$, $Q \circ y(u, v) = 0$, and $R \circ z(u, v) = \sin^2 u \cos^2 v$. (iv) $\frac{\partial(x,y)}{\partial(u,v)} = \sin u \cos u, \frac{\partial(y,z)}{\partial(u,v)} = 0$, and $\qquad \frac{\partial(z,x)}{\partial(u,v)} = 0.$ (v) $\int_{u_0}^{u_1} \int_{v_0}^{v_1} P \circ T(u,v)\frac{\partial(y,z)}{\partial(u,v)} + Q \circ T(u,v)\frac{\partial(z,x)}{\partial(u,v)} + R \circ$

$T(u,v)\frac{\partial(x,y)}{\partial(u,v)} \, dv \, du = \int_0^{2\pi} \int_0^{\frac{\pi}{2}} \sin^3 u\cos^2 v\cos u \, du \, dv = \frac{\pi}{4}.$

10.4. GREEN THEOREM IN $\mathbb{G}_2$

Definition 10.5. Let $D \subset \mathbb{R}^2$ be a region and ∂D be its closed counter-clockwise orientation boundary defined by $c(t) = (x(t), y(t))$. Let the 1-form function $w = P(x,y) + Q(x,y)$ be on some open set K in $\mathbb{R}^2$ that contains D [25]. Then

$$\int_{\partial D} P \circ c(t) \frac{x(t)}{dt} + Q \circ c(t) \frac{y(t)}{dt} = \iint_D \left(\frac{\partial Q}{\partial x} - \frac{\partial P}{\partial y}\right) dx \wedge dy \qquad (10.8)$$

Example 10.7. Verify Green's theorem over $\int_D xydx + x^2y^2dy$, where c is the triangle with vertices $(0,0)$, $(1,0)$, and $(0,1)$.

Solution 10.7.

$$\iint_D \left(\frac{\partial Q}{\partial x} - \frac{\partial P}{\partial y}\right) dydx \qquad = \int_0^1 \int_0^{1-x} 2xy^2 - xdy \, dx$$

$$= \int_0^1 \frac{-2x^4+6x^3-3x^2-x}{3}$$

$$= -\frac{2}{15}$$

$$\int_{\partial D} P \circ c(t) \frac{x(t)}{dt} + Q \circ c(t) \frac{y(t)}{dt} \quad = \int_{\partial D_1} P \cdot c_1(t)\frac{dx}{dt} + Q \cdot c_1(t)\frac{dy}{dt} \qquad (10.9)$$

$$+ \int_{\partial D_2} P \cdot c_2(t)\frac{dx}{dt} + Q \cdot c_2(t)\frac{dy}{dt}$$

$$+ \int_{\partial D_3} P \cdot c_3(t)\frac{dx}{dt} + Q \cdot c_3(t)\frac{dy}{dt}$$

$$= 0 - \frac{2}{15} + 0$$

$$= -\frac{2}{15}$$

where $c_1(t) = (t,0), t \in [0,1],$ $c_2(t) = (1-t,t), t \in [0,1],$ and $c_3(t) = (0,t), t \in [1,0].$

$$\int_{\partial D_1} P \cdot c_1(t)\frac{dx}{dt} + Q \cdot c_1(t)\frac{dy}{dt} = \int_0^1 xy(0) + x^2y^2(1)dt$$
$$= \int_0^1 0(1) + 0(0)dt$$
$$= 0$$

$$\int_{\partial D_2} P \cdot c_2(t)\frac{dx}{dt} + Q \cdot c_2(t)\frac{dy}{dt} = \int_0^1 xy(-1) + x^2y^2(1)dt$$
$$= \int_0^1 -t(1-t) + (1-t)^2(t^2)dt \qquad (10.10)$$
$$= -\frac{2}{15}$$

$$\int_{\partial D_3} P \cdot c_3(t)\frac{dx}{dt} + Q \cdot c_3(t)\frac{dy}{dt} = \int_1^0 xy(1) + x^2y^2(0)dt$$
$$= \int_1^0 0(0) + 0(1)dt$$
$$= 0$$

10.5. STOKES' THEOREM IN $\mathbb{G}_3$

Definition 10.6. Let $S \subset \mathbb{R}^3$ be a closed region and ∂S be its counter-clockwise orientation surface defined by $c(t) = (x(t), y(t), z(t))$. Let the 1-form function $w = P(x, y, z) + Q(x, y, z) + R(x, y, z)$ be on some open set K in $\mathbb{R}^3$ that contains S region [25].

$$\int_{\partial S} P \circ c(t)\frac{x(t)}{dt} + Q \circ c(t)\frac{y(t)}{dt} + R \circ c(t)\frac{z(t)}{dt} =$$
$$\iint_S \left(\frac{\partial R}{\partial y} - \frac{\partial Q}{\partial z}\right)dy \wedge dz + \left(\frac{\partial P}{\partial z} - \frac{\partial R}{\partial x}\right)dz \wedge dx + \left(\frac{\partial Q}{\partial x} - \frac{\partial P}{\partial y}\right)dx \wedge dy \qquad (10.11)$$

where $dy \wedge dz = \frac{\partial(T_y, T_z)}{\partial(u,v)}$, $dz \wedge dx = \frac{\partial(T_z, T_x)}{\partial(u,v)}$, and $dx \wedge dy = \frac{\partial(T_x, T_y)}{\partial(u,v)}$.

Example 10.8. Verify Stokes's theorem over $\int_D x^2 dx + y^2 dy + z^2 dz$, where c is defined by $c(t) = (\cos t, \sin t, 1 - \cos t - \sin t)$

Solution 10.8. If $T(\theta, r) = (r\cos t, r\sin t, 1 - r\cos t - r\sin t)$,

$$\frac{\partial(T_y, T_z)}{\partial(\theta, r)} = \begin{vmatrix} r\cos\theta & \sin\theta \\ r\sin\theta & -r\cos\theta \end{vmatrix} \qquad (10.12)$$

$$\frac{\partial(T_z,T_x)}{\partial(\theta,r)} = \begin{vmatrix} r\sin\theta & -r\cos\theta \\ -r\sin\theta & \sin\theta \end{vmatrix} \tag{10.13}$$

$$\frac{\partial(T_x,T_y)}{\partial(\theta,r)} = \begin{vmatrix} -r\sin\theta & \sin\theta \\ r\cos\theta & \sin\theta \end{vmatrix} \tag{10.14}$$

$$
\begin{aligned}
\iint_S dw &= \int_0^1 \int_0^{2\pi} \left(\frac{\partial R}{\partial y} - \frac{\partial Q}{\partial z}\right) dydz + \left(\frac{\partial P}{\partial z} - \frac{\partial R}{\partial x}\right) dzdx + \left(\frac{\partial Q}{\partial x} - \frac{\partial P}{\partial y}\right) dxdy \\
&= \int_0^1 \int_0^{2\pi} \left(\frac{\partial R}{\partial y} - \frac{\partial Q}{\partial z}\right) \circ T(\theta,r) \frac{\partial(T_y,T_z)}{\partial(\theta,r)} \\
&+ \left(\frac{\partial P}{\partial z} - \frac{\partial R}{\partial x}\right) \circ T(\theta,r) \frac{\partial(T_z,T_x)}{\partial(\theta,r)} \\
&+ \left(\frac{\partial Q}{\partial x} - \frac{\partial P}{\partial y}\right) \circ T(\theta,r) \frac{\partial(T_x,T_y)}{\partial(\theta,r)} \\
&= \int_0^1 \int_0^{2\pi} (0)\frac{\partial(T_y,T_z)}{\partial(\theta,r)} + (0)\frac{\partial(T_z,T_x)}{\partial(\theta,r)} + (0)\frac{\partial(T_x,T_y)}{\partial(\theta,r)} \, d\theta dr \\
&= 0
\end{aligned}
\tag{10.15}
$$

$$
\begin{aligned}
\int_{\partial D} P \cdot c(t)\frac{dx}{dt} + Q \cdot c(t)\frac{dy}{dt} + R \cdot c(t)\frac{dy}{dt} &= \int_0^{2\pi} (\cos t)^2(-\sin t) \\
&+ (\sin t)^2(\cos t) \\
&+ (1 - \cos t - \sin t)^2(\sin t - \cos t)dt \\
&= 0
\end{aligned}
\tag{10.16}
$$

10.6. GAUSS' THEOREM IN $\mathbb{G}_3$

Definition 10.7. Let $\Omega \subset \mathbb{R}^3$ be a closed and solid region, and $\partial\Omega$ be its counter-clockwise orientation surface. Let the 2-form function $\eta = P(x,y,z)dydz + Q(x,y,z)dzdx + R(x,y,z)dxdy$ be on some open set K in $\mathbb{R}^3$ that contains the Ω region [25].

$$
\begin{aligned}
\iint_{\partial\Omega} P(x,y,z)dy \wedge dz + Q(x,y,z)dz \wedge dx + R(x,y,z)dx \wedge dy &= \\
\iiint_\Omega \left(\frac{\partial P}{\partial x} + \frac{\partial Q}{\partial y} + \frac{\partial R}{\partial z}\right) dz\, dy\, dx
\end{aligned}
\tag{10.17}
$$

Example 10.9. Verify Gauss' theorem over $\int_V xdydz + ydzdx + zdxdy$, where c is defined by $T(\theta,\phi) = (\cos\theta\sin\phi, \sin\theta\sin\phi, \cos\phi)$.

Solution 10.9.

$$\int_{\Omega} = \iiint_{\Omega} \left(\frac{\partial P}{\partial x} + \frac{\partial Q}{\partial y} + \frac{\partial R}{\partial z} \right) dz\, dy\, dx$$

$$= \int_{-1}^{-1} \int_{-\sqrt{1-x^2}}^{\sqrt{1-x^2}} \int_{-\sqrt{1-x^2-y^2}}^{\sqrt{1-x^2-y^2}} 3\, dz\, dy\, dx \qquad (10.18)$$

$$= 4\pi$$

$$\iint_{\partial\Omega} d\Omega = \int_0^{\pi} \int_0^{2\pi} P \circ T \frac{\partial(T_y, T_z)}{\partial(\theta,\phi)} + Q \circ T \frac{\partial(T_z, T_x)}{\partial(\theta,\phi)} + R \circ T \frac{\partial(T_x, T_y)}{\partial(\theta,\phi)} d\theta\, d\phi$$

$$= \int_0^{\pi} \int_0^{2\pi} \cos\theta\sin\phi \frac{\partial(T_y, T_z)}{\partial(\theta,\phi)} + \sin\theta\sin\phi \frac{\partial(T_z, T_x)}{\partial(\theta,\phi)} + \cos\phi \frac{\partial(T_x, T_y)}{\partial(\theta,\phi)}$$

$$= \int_0^{\pi} \int_0^{2\pi} -\sin^3\phi - \cos^2\phi\sin^2\theta\sin\phi - \cos^2\theta\sin\phi\cos^2\phi \, d\theta d\phi \qquad (10.19)$$

$$= \int_0^{\pi} \int_0^{2\pi} -\sin^3\phi - \cos^2\phi\sin\phi \, d\theta d\phi$$

$$= -4\pi$$

Note 10.1. The sign depends on the orientation.

$$\frac{\partial(T_y, T_z)}{\partial(\theta,r)} = \begin{vmatrix} \cos\theta\cos\phi & \sin\theta\cos\phi \\ 0 & -\sin\phi \end{vmatrix} \qquad (10.20)$$

$$= -\cos\theta\sin^2\phi$$

$$\frac{\partial(T_z, T_x)}{\partial(\theta,r)} = \begin{vmatrix} 0 & -\sin\phi \\ -\sin\theta\sin\phi & \cos\theta\cos\phi \end{vmatrix} \qquad (10.21)$$

$$= -\sin^2\phi$$

$$\frac{\partial(T_x, T_y)}{\partial(\theta,r)} = \begin{vmatrix} -\sin\theta\sin\phi & \cos\theta\cos\phi \\ \cos\theta\sin\phi & \sin\theta\cos\phi \end{vmatrix} \qquad (10.22)$$

$$= -\sin^2\theta\sin\phi\cos\phi - \cos^2\theta\sin\phi\cos\phi$$

10.7. EXERCISES

Exercise 10.1. Let $w = \frac{-y}{x^2+y^2} dx + \frac{x}{x^2+y^2}$?. Compute the line integral over the unit circle oriented counter-clockwise.

Exercise 10.2. Let the function $w = xdxdy$ be in space ?, and the oriented surface S representing the unit circle with centre at point $(0,0,0)$. (i) Determine $P(x,y,z)$, $Q(x,y,z)$, and $R(x,y,z)$. (ii) Determine $x(u,v)$, $y(u,v)$, and $z(u,v)$.

(iii) Determine $P \circ T(u, v)$, $Q \circ T(u, v)$, and $R \circ T(u, v)$. (iv) Determine $\frac{\partial(x,y)}{\partial(u,v)}$, $\frac{\partial(y,z)}{\partial(u,v)}$, and $\frac{\partial(z,x)}{\partial(u,v)}$. (v) Compute the surface integral over the surface $T(u, v)$.

Exercise 10.3. Verify Green's theorem over $\int_D ydx + x^2 dy$, where c is the square with vertices $(0,0)$, $(1,0)$, $(0,1)$, and $(1,1)$.

Exercise 10.4. Verify Stokes' theorem over $\int_D xdx + ydy + zdz$, where c is defined by $c(t) = (\cos t, \sin t, 1 - \cos t - \sin t)$.

Exercise 10.5. Verify Gauss' theorem over $\int_V x^2 dydz + y^2 dzdx + zdxdy$, where c is defined by $T(\theta, \phi) = (\cos\theta\sin\phi, \sin\theta\sin\phi, \cos\phi)$.

CHAPTER 11

Solutions for Chapters

Abstract: This chapter provides the complete solution to all the exercises pointed out at the end of each chapter of this book. The solutions not only indicate the final result, but the respective procedures are exhibited. It is recommended in all cases that the reader review the solutions to all the exercises.

Keywords: Bordered Hessian matrix, Cartesian coordinates, Critical points, Cross product, Cylindrical coordinates, Differential forms, Directional derivatives, Divergence, Double riemann integral, Exterior derivative, Exterior product, Gauss' theorem, Geometric product, Graph of vector valued-function, Grassmann algebra, Green's theorem, Hessian matrix, L'Hospital's rule, Implicit function theorem, Improper integral in space, Improper integral on a plane, Inner product, Inverse function theorem, Jacobian determinant, Line integral of vector functions, Maps in $\mathbb{R}$, Maps in $\mathbb{R}^2$, Maps in $\mathbb{R}^3$, Partial derivatives, Stokes' theorem, Surface integral of vector functions, Taylor's theorem, Triple riemann integral.

SOLUTIONS FOR CHAPTER 1

Solution 1.1. *Proof.* Let p, $p_0 \in \mathbb{R}^3$, $p_0 = (x_0, y_0, z_0)$, lie on the plane P, $p = (x, y, z)$ be at any point in space, and a normal vector $n = (A, B, C)$ be in that plane. Then p lies on the plane P if vector $p - p_0$ is perpendicular to vector n, that is, $(p - p_0) \cdot n = 0$. So the equation of the plane P is $A(x - x_0) + B(y - y_0) + C(z - z_0) = 0$

Solution 1.2. Let vector $c = d + v \Rightarrow d = (3,1,4)$ and vector $v = b - a = (2,0,-1) - (-1,2,3) \Rightarrow v = (3,-2,-4)$. The point c is $(3,1,4) + (3,-2,-4) = (6,-1,0)$ Fig. (**1.10**).

Solution 1.3. First we solve for x in $21x + 80y = 13$, $y = \frac{13-21x}{80}$. Now we determine two points. If $x = 1 \Rightarrow y = -0.1$ and $x = 2 \Rightarrow y = -0.3625$. The direction of vector v is $(1,-0.1) - (2,-0.3625) = (-1,0.2625) \Rightarrow \frac{v}{||v||} = (-0.9673,0.2539)$. The vector w is $14\frac{v}{||v||} = (-13.5422,3.5548)$.

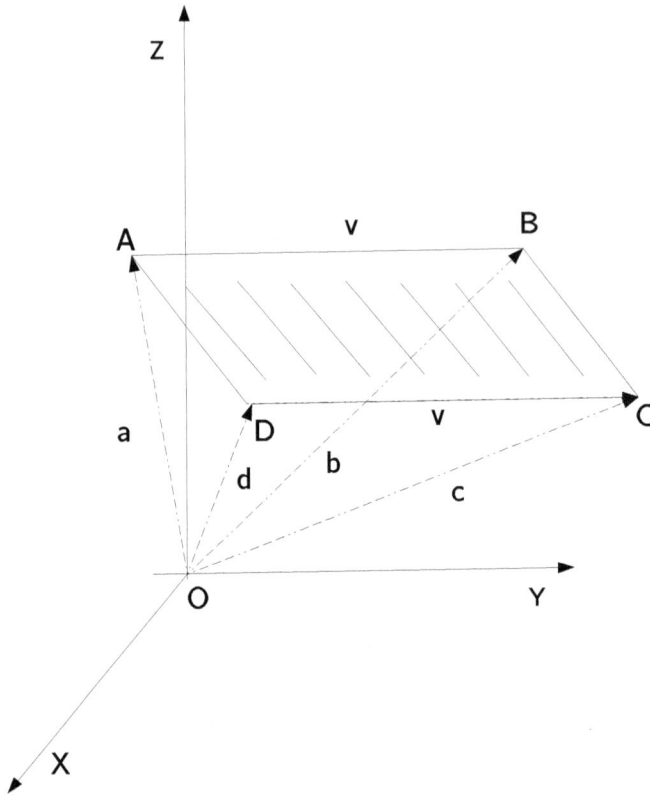

Fig. (1.10). Vector c is represented by the segment CO, where $CO = DO + CD$.

Solution 1.4. Let vector $a = \overleftarrow{PQ} = Q - P = (1,0,1) - (1,1,0) = (0,-1,1)$ and vector $b = \overleftarrow{PR} = R - P = (0,1,1) - (1,1,0) = (-1,0,1)$. The area of the triangle is $\frac{1}{2}||a \times b|| = \frac{1}{2}||(-1,-1,-1)|| = \frac{\sqrt{3}}{2}$.

Solution 1.5. $u \times v = (2c, c, b - 2a) \Rightarrow ||(2c, c, b - 2a)|| = \sqrt{5c^2 + (b - 2a)^2}$. Thus, the area is $\sqrt{5c^2 + (b - 2a)^2}$.

Solution 1.6. The magnitude of segment $\overleftarrow{PR}$ is the distance required (Fig. **1.11**). From the equation plane, the normal vector n is $(2,3,-1)$. Then $L = p + tn = (1,4,1) + t(2,3,-1) \Rightarrow x = 1 + 2t, y = 4 + 3t$, and $z = 1 - t$. $2(1 - 2t) + 3(4 + 3t) - (1 - t) = -1$; $t = -1$. The intercept of the line in the plane is the point R, $(1 + 2(-1), 4 + 3(-1), 1 - (-1)) = (-1,1,0)$. Hence, the segment $\overleftarrow{PR}$ is $(-1,1,0) - (1,4,1) = (-2,-3,-1)$ and its length $||(-2,-3,-1)|| = \sqrt{14}$. [0.7].

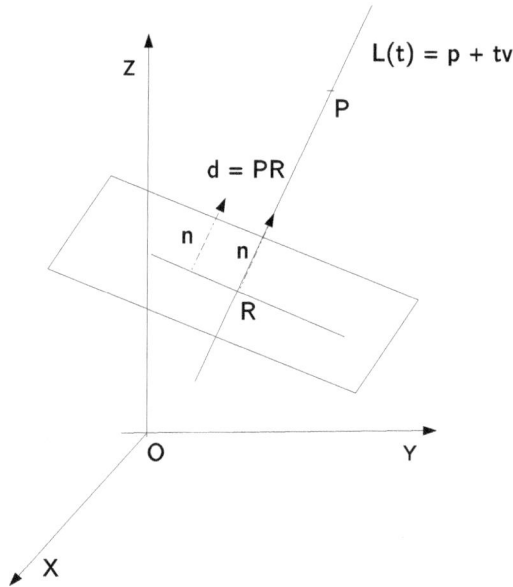

Fig. (1.11). The segment PR is the distance required.

Solution 1.7. A normal vector n to $x + y - z = 1$ and $2x - 3y + 4z = 5$ is $n = n_1 \times n_2 = (1,1,-1) \times (2,-3,4) = (1,-6,-5)$, n_1, n_2 are normal vectors to the plane. Hence, P is $[(x,y,z) - P] \cdot n = 0 \Leftrightarrow [(x,y,z) - (1,0,-2)] \cdot n = 0 \Leftrightarrow (x-1, y, z-2) \cdot (1,-6,-5) = 0 \Leftrightarrow x - 6y - 5z = 11$. $P(s,t) = (1,0,-2) + s(0,0,-\frac{11}{5}) + t(1,1,\frac{6}{5})$.

Solution 1.8. Let the vector $a = Q - P = (-1,-3,3) - (1,-5,2) = (-2,2,1)$ and vector $v = R - P = (-3,-1,5) - (1,-5,2) = (-4,4,3)$. If vectors a and b are collinear, it means that $a = \alpha b$, but $\nexists\ \alpha \in \mathbb{R}$, such that $\alpha a = b$. The points P, Q, and R are not collinear.

SOLUTIONS FOR CHAPTER 2

Solution 2.1. (i) Let $\lim_{(x,y)\to(0,0)} \frac{xy}{x+y}$, if it approaches $(0,0)$ through $(x, -x^2 - x) \Rightarrow \lim_{(x,-x^2-x)\to(0,0)} \frac{xy}{x+y} = \lim_{x\to 0} \frac{-x^3-x^2}{x-x^2-x} = 1$. but it approaches $(0,0)$ through $(x, x^2 - x) \Rightarrow \lim_{(x,x^2-x)\to(0,0)} \frac{xy}{x+y} = \lim_{x\to 0} \frac{x^3-x^2}{x+x^2-x} = -1$. Then, the limit of f does not exist. (ii) $\lim_{(x,y)\to(5,1)} \frac{xy}{x+y} = \frac{5}{6}$.

Solution 2.2. Let $\lim_{(x,y)\to(0,0)} \frac{x^3y}{x^6+y^2}$, if it approaches $(0,0)$ through $(x,x) \Rightarrow$

$\lim_{(x,-x)\to(0,0)} \frac{x^3x}{x^6+x^2} = \lim_{x\to0} \frac{x^2x^2}{x^2(x^4+1)} = 0$, but it approaches $(0,0)$ through

$(x,x^3) \Rightarrow \lim_{(x,x^3)\to(0,0)} \frac{x^3x^3}{x^6+x^6} = \lim_{x\to0} \frac{x^6}{2x^6} = \frac{1}{2}$. Then, the limit of f does not exist.

Solution 2.3. Let the unit vector $\frac{v}{||v||} = \frac{1}{\sqrt{20}}(3,-3,2)$. $Df_v = grad f_{x_0=(3,-3,2)} \cdot$
$\frac{v}{||v||} = (yz^3, xz^3, 3xyz^2)f_{x_0=(3,-3,2)} \cdot \frac{v}{||v||} = (-24,24,-108) \cdot (\frac{3}{\sqrt{20}}, -\frac{3}{\sqrt{20}}, \frac{2}{\sqrt{20}}) =$
$-\frac{180}{\sqrt{5}}$.

Solution 2.4. Let the function $r: \mathbb{R} \to \mathbb{R}^3, (2t+1, 3t, 1-t)$ and the point $t_0 = 1$ then

$$D(d \circ p(t)) = \begin{pmatrix} \frac{\partial 2t+1}{\partial t} \\ \frac{\partial 3t}{\partial t} \\ \frac{\partial 1-t}{\partial t} \end{pmatrix}_{t_0=1} = \begin{pmatrix} 2 \\ 3 \\ -1 \end{pmatrix}.$$

Solution 2.5. It is not possible to obtain the function $f \circ g$, because the image of function g is not in $\mathbb{R}^3$.

Solution 2.6. Let $h: \mathbb{R}^2 \to \mathbb{R}$, and $p: \mathbb{R} \to \mathbb{R}^2$. (i) The function $h \circ p: \mathbb{R} \to \mathbb{R}, h \circ p = h(p(t)) = h(t, \frac{t}{2}) = \frac{t^2}{2}$.

(ii) $\qquad D\left(h \circ p(t_0)\right) = \left(\frac{\partial xy}{\partial x} \frac{\partial xy}{\partial y}\right)_{y_0} \begin{pmatrix} \frac{\partial t}{\partial t} \\ \frac{\partial \frac{t}{2}}{\partial t} \end{pmatrix}_{t_0} = (y \; x)_{y_0} \begin{pmatrix} 1 \\ \frac{1}{2} \end{pmatrix}.$

Solution 2.7. Let the function $h \circ p: \mathbb{R}^3 \to \mathbb{R}^3, (xy^2, e^x \sin xy, x\ln yx)$, its $Df(x,y,z)$ at point $x_0 = (2,1,2)$ is

$$Df(x,y,z) = \begin{pmatrix} \dfrac{\partial xy^2}{\partial x} & \dfrac{\partial xy^2}{\partial y} & \dfrac{\partial xy^2}{\partial z} \\[2ex] \dfrac{\partial e^x \sin xy}{\partial x} & \dfrac{\partial e^x \sin xy}{\partial y} & \dfrac{\partial e^x \sin xy}{\partial z} \\[2ex] \dfrac{\partial x \ln yx}{\partial x} & \dfrac{\partial x \ln yx}{\partial y} & \dfrac{\partial x \ln yx}{\partial z} \end{pmatrix}_{x_0}$$

$$Df(x,y,z) = \begin{pmatrix} y^2 & 2xy & 0 \\[2ex] ye^x \cos xy + e^x \sin xy & xe^x \cos xy & 0 \\[2ex] 1 + \ln xy & \dfrac{x}{y} & 0 \end{pmatrix}_{x_0}$$

$$= \begin{pmatrix} 1 & 4 & 0 \\[2ex] 2e^2 \cos 2 + e^2 \sin 2 & 2e^2 \cos 2 & 0 \\[2ex] 1 + \ln 2 & 2 & 0 \end{pmatrix}$$

SOLUTIONS FOR CHAPTER 3

Solution 3.1. Let $\frac{\partial T}{\partial t} = -ke^{-kt}\cos x$ and $k\frac{\partial^2 T}{\partial x} = -e^{-kt}\cos x$. The equation is verified.

Solution 3.2. Let $\frac{\partial f}{\partial x} = 2x + 1 + y = 0$ and $\frac{\partial f}{\partial y} = 2y + 1 + x = 0 \Rightarrow x = y = -\frac{1}{3}$.
There is a unique critical point $(-\frac{1}{3}, -\frac{1}{3})$.

Solution 3.3. Let $(1)\frac{\partial f}{\partial x} = \frac{2x}{1+x^2+y^2} - \frac{2x}{1+x^4} = 0$ and $(2)\frac{\partial f}{\partial y} = \frac{2y}{1+x^2+y^2} = 0$. From (2), $y = 0$. From (2), $\frac{2x}{1+x^4} = 0, \Rightarrow x^3(x^2 - 1) = 0$. There are three critical points $(0,0)$, $(-1,0)$, and $(x - 1,0)$.

Solution 3.4. Let $(1)\frac{\partial f}{\partial x} = 3x^2 - 6y = 0$ and $(2)\frac{\partial f}{\partial y} = -6x + 3y^2 = 0$. From (1) and (2) the critical points are $(0,0)$ and $(2,2)$. The method is not conclusive at point $(0,0)$. The point $(2,2)$ is minimum.

$$H_1(f(x)) = \left[\frac{\partial^2 x^3 - 6xy + y^3}{\partial x\, \partial x}\right]_{x_0=(0,0)} = [0] = 0$$

$$H_2(f(x)) = \begin{vmatrix} \dfrac{\partial^2 x^3 - 6xy + y^3}{\partial x\, \partial x} & \dfrac{\partial^2 x^3 - 6xy + y^3}{\partial x\, \partial y} \\[2mm] \dfrac{\partial^2 x^3 - 6xy + y^3}{\partial y\, \partial x} & \dfrac{\partial^2 x^3 - 6xy + y^3}{\partial y\, \partial y} \end{vmatrix}_{x_0=(0,0)}$$

$$= \begin{bmatrix} 0 & -6 \\ -6 & 0 \end{bmatrix}_{x_0=(0,0)} < 0$$

$$H_1(f(x)) = \left[\frac{\partial^2 x^3 - 6xy + y^3}{\partial x\, \partial x}\right]_{x_0=(2,2)} = [12] > 0$$

$$H_2(f(x)) = \begin{vmatrix} \dfrac{\partial^2 x^3 - 6xy + y^3}{\partial x\, \partial x} & \dfrac{\partial^2 x^3 - 6xy + y^3}{\partial x\, \partial y} \\[2mm] \dfrac{\partial^2 x^3 - 6xy + y^3}{\partial y\, \partial x} & \dfrac{\partial^2 x^3 - 6xy + y^3}{\partial y\, \partial y} \end{vmatrix}_{x_0=(2,2)}$$

$$= \begin{bmatrix} 12 & -6 \\ -6 & 12 \end{bmatrix}_{x_0=(2,2)} > 0$$

Solution 3.5. Let $\frac{\partial f}{\partial x} = 0$ and $\frac{\partial f}{\partial y} = 0$. The critical points are the points in $\mathbb{R}^2$.

Solution 3.6. (i) By $\nabla f(x) = (0,0)$, we obtain only one critical point $(0,0)$.

$$\nabla f(x) = \left(\frac{\partial x^3 + y^3}{\partial x}, \frac{\partial x^3 + y^3}{\partial y}\right) = (3x^2, 3y^2) = (0,0)$$

(ii) The method cannot be applied to qualify this point.

$$H_1(f(x)) = \left[\frac{\partial^2 x^3 + y^3}{\partial x\, \partial x}\right]_{x_0=(0,0)} = [6x]_{x_0=(0,0)} = 0 = 0$$

$$H_2(f(x)) = \begin{vmatrix} \dfrac{\partial^2 x^3 + y^3}{\partial x\, \partial x} & \dfrac{\partial^2 x^3 y^3}{\partial x\, \partial y} \\[2em] \dfrac{\partial^2 x^3 + y^3}{\partial y\, \partial x} & \dfrac{\partial^2 x^3 + y^3}{\partial y\, \partial y} \end{vmatrix}_{x_0=(0,0)} = \begin{bmatrix} 0 & 0 \\ 0 & 0 \end{bmatrix}_{x_0=(0,0)} = 0$$

$$= 0$$

Solution 3.7. Since $\nabla f(x_0) = \lambda \nabla g(x_0) \Rightarrow (1,1) = \lambda(2x, 2y)$, is restricted to S, it implies that $x = \frac{1}{2\sqrt{\lambda}}$ and $y = \frac{1}{2\sqrt{\lambda}} \Rightarrow \lambda = \pm\frac{1}{\sqrt{2}}$. The critical points are $(\lambda, x, y) = \pm\left(\frac{1}{\sqrt{2}}, \frac{\sqrt{2}}{2}, \frac{\sqrt{2}}{2}\right)$.

Solution 3.8. (i) Since $F(1,2,2) = 0$, F is class C^1 and $\frac{\partial F}{\partial z}(1,2,2) = 4 > 0$. There is a neighbourhood of point $x = (1,2)$, and there exists a unique solution $z = f(x, y)$ derivable with a continuous derivative such that $f(1,2) = 2$ So, this is the solution to F. (ii) $\frac{\partial f}{\partial x}_{(x_0,y_0)} = -\frac{2x}{2z}_{(1,2)} = -\frac{1}{2}$ and $\frac{\partial f}{\partial y}_{(x_0,y_0)} = -\frac{2y}{2z}_{(x_0,y_0)} = -1$.

(iii) $f(x, y) = f(x_0, y_0) + \frac{\partial f}{\partial x}_{(x_0,y_0)}(x - x_0) + \frac{\partial f}{\partial y}_{(x_0,y_0)}(y - y_0) +$

$\frac{1}{2}\left[\frac{\partial^2 f}{\partial x\, \partial x}_{(x_0,y_0)}(x - x_0)^2 + \frac{\partial^2 f}{\partial x\, \partial y}_{(x_0,y_0)}(x - x_0)(y - y_0) + \right.$

$\left. \frac{\partial^2 f}{\partial y\, \partial x}_{(x_0,y_0)}(x - x_0)(y - y_0) + \frac{\partial^2 f}{\partial y\, \partial y}_{(x_0,y_0)}(y - y_0)^2\right] = 2 - \frac{1}{2}(x - 1) -$

$\frac{1}{2}(y - 2)$.

Particularly, $f(1,2) = 2$.

Solution 3.9. (i) Let $f(x) = \sin x \rightarrow Jf(0) = |\sin x|_{x=0} = \cos 0 = 1 \neq 0$, then there exists a function f^{-1} of class C^{-1} to function f. (ii) Since $\frac{\partial f}{\partial x}_{x_0=0} = 1$ and

$\frac{\partial f^{-1}}{\partial y}_{y_0} = \frac{1}{\frac{\partial f}{\partial x}_{x_0}}$

$= \frac{1}{1} = 1$. (iii) $f^{-1}(x) = f(y) = f(y_0) + \frac{\partial f}{\partial y}_{(y_0)}(y - y_0) + \frac{1}{2}\frac{\partial^2 f}{\partial y\, \partial y}_{(y_0)}$

$(y - y_0)^2 + \frac{1}{6} \frac{\partial^3 f}{\partial y \, \partial y \, \partial y}_{(y_0)} (y - y_0)^3 = 1 + 1(y - 1) = 1 + 1y - 1^2$. Particularly, $f(0) = 1$.

SOLUTIONS FOR CHAPTER 4

Solution 4.1. The vector field of the vector-valued function F (Fig. **4.7**).

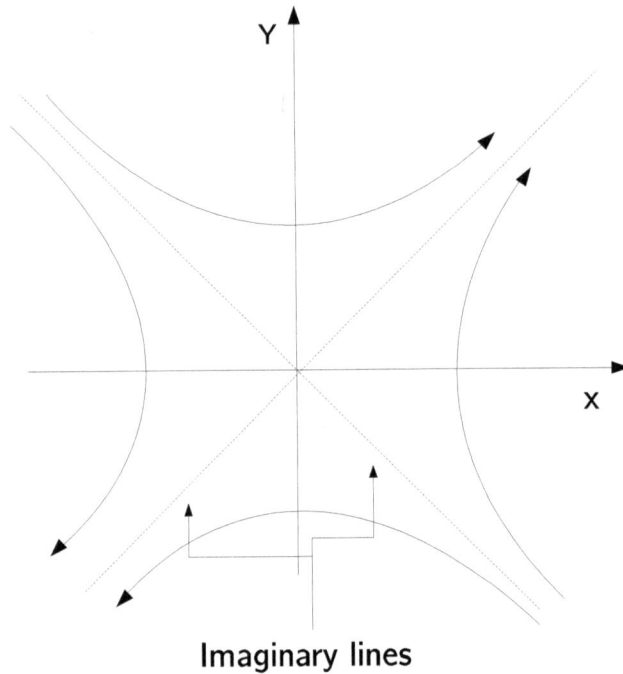

Imaginary lines

Fig. (4.7). Vector field of $F(x, y) = (y, x)$.

Solution 4.2. (i) The divergence of the vector-valued function F is $\mathrm{div} F\,(2,3,1) = \frac{\partial x}{\partial x} + \frac{\partial y}{\partial y} + \frac{\partial z}{\partial z} (2,3,1) = 3$. (ii) The rotational of function F is $\mathrm{rot}\, F(7,2,1) = (0,0,0)$. (iii) Suppose the flow line $c(t) = (1,1,1) + t(1,1,1), t \in \mathbb{R}$. The $c'(t) = (1,1,1) = F \circ C(t) = (t,t,t)_{t=1}$. Then $C(t)$ is a flow line of the vector-valued function F.

Solution 4.3. If $\mathrm{rot}\, F = (\frac{\partial F_3}{\partial y} - \frac{\partial F_2}{\partial z}, \ -\frac{\partial F_3}{\partial x} - \frac{\partial F_1}{\partial z}, \ \frac{\partial F_2}{\partial x} - \frac{\partial F_1}{\partial y})$. So $\mathrm{div}\,(\mathrm{rot})\, F = \frac{\partial}{\partial x}(\frac{\partial F_3}{\partial y} - \frac{\partial F_2}{\partial z}) - \frac{\partial}{\partial y}(\frac{\partial F_3}{\partial x} - \frac{\partial F_1}{\partial z}) + \frac{\partial}{\partial z}(\frac{\partial F_2}{\partial x} - \frac{\partial F_1}{\partial y}) = 0.$

Solution 4.4. (i) The divergence of the vector-valued function F is $\text{div} F$ $(2,3) =$ $\frac{\partial -y}{\partial x} + \frac{\partial x}{\partial y} + \frac{\partial 0}{\partial z}(2,3) = 0$. (ii) The rotational of the vector-valued function F is $rot\, F(2,1) = (0,0,2)$. (iii) Suppose the flow line $c(t) = (\cos t, \sin t)\, t \in [0,2\pi]$. The $c'(t) = (-\sin t, \cos t) = F \circ C(t) = (-\sin t, \cos t)$. Then $C(t)$ is a flow line of the vector-valued function F.

Solution 4.5. The vector function $F(x,y) = (y,x)$ is a conservative vector field because its potential function $f(x,y) = xy$, $F(x,y) = \nabla f = (y,x)$; but the vector function $F(x,y) = (-y,x)$ is **not** a conservative vector field, because there is not a function f such that $F = \nabla f$.

Solution 4.6. We need $l(t) = (x(t), y(t))l'(t) = F(l(t))$ and $(x'(t), y'(t)) = (1, y(t))$. Then $x'(t) = 1$ and $y'(t) = y(t)$. $\int x'(t)\, dt = \int dt \rightarrow x(t) = t + c_1$. $y'(t) - y(t) = 0, \frac{dy}{dt} = y(t) \Rightarrow \int \frac{dy}{y} \Rightarrow \int dt = \ln|y| = t + c_2.|y| = e^{t+c_2}$ or $y = \pm C_3 e^t$. So, $l(t) = (t, Ce^t)$ meets the restrictions.

Solution 4.7. (i) $rot(\nabla f)$ is a vector. (ii) $div(rot f) = 0$. (iii) $\nabla \cdot (\nabla \times F)$ is a scalar.

SOLUTIONS FOR CHAPTER 5

Solution 5.1. (i) $\int_1^2 x^2 \sqrt{x^3}\, dx = \int_1^2 x^{\frac{7}{2}}\, dx = \left[\frac{2x^{\frac{9}{2}}}{9} \right]_1^2 = 4.8060$. (ii) Let $T(u) =$ $u^{\frac{1}{3}}$, $u \in [1,8]$. (iii) If $T(u) = u^{\frac{1}{3}}$, then $J(T(u)) = \frac{1}{3u^{\frac{2}{3}}}$; thus $\int_1^2 x^2 \sqrt{x^3}\, dx = \int_1^8 f \circ$ $T(u) \; |J(T(u)|du = \frac{1}{3} \int_1^8 \sqrt{u}\, du = \int_1^8 \sqrt{u}\, du = 4.8060$. (iv) If $u = x^3 \Rightarrow du =$ $3x^2\, dx \Rightarrow dx = \frac{du}{3x^2} \Rightarrow \int_1^8 x^2 \sqrt{u}\; \frac{du}{3x^2} = \frac{2}{9}(\sqrt{o^3 - 1}) = 4.8060$. (v) The closed interval $[1,8]$ is transformed into the close interval $[1,2]$ by $T(u) = u^{\frac{1}{3}}$ Fig. (**5.16**).

D* space

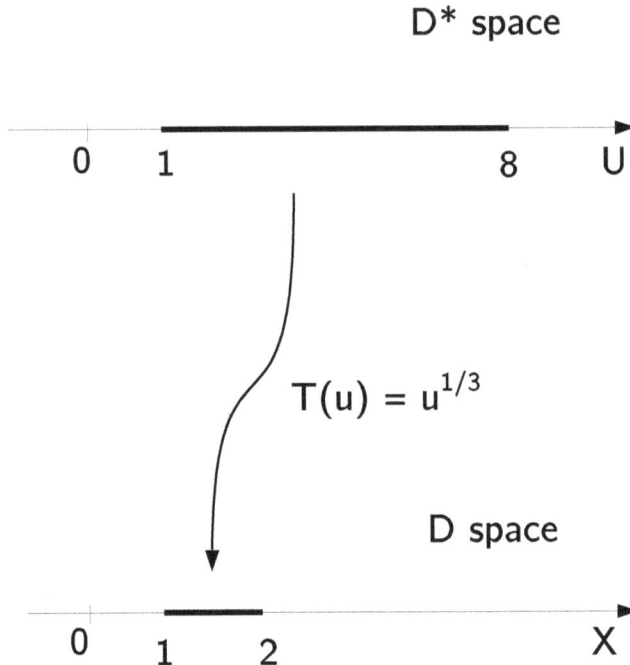

Fig. (5.10). The T mapping from D^* space to D space.

Solution 5.2. (i) $\iint_D x^2 + y^2 \, dy \, dx = \int_{-1}^{1} \int_{-\sqrt{1-x^2}}^{\sqrt{1-x^2}} x^2 + y^2 \, dy \, dx = \int_{-1}^{1} \frac{2}{3}(4x^2 + 1)$

$\sqrt{x^2 + 1} \, dx = \left[\frac{2}{3}x(x^2 + 1)^{\frac{3}{2}}\right]_{-1}^{1} = \frac{8\sqrt{2}}{3}$. (ii) Let $T(r, \theta) = (r\cos\theta, r\sin\theta)$ (r, θ)

$\subset [1,0] \times [0,2\pi]$. (iii) $J(T(r, \theta)) = r$; $\int_{-1}^{1} \int_{-\sqrt{x^2+1}}^{\sqrt{x^2+1}} x^2 + y^2 \, dy \, dx =$

$\int_{0}^{2\pi} \int_{0}^{1} f \circ T(r, \theta) \, |J(T(r, \theta)| dr \, d\theta = 2\pi \int_{0}^{1} r^3 \, dr = \frac{\pi}{2}$. (iv) The closed region $[0,1] \times [0,2\pi]$ is transformed into the close unit circle by $T(r, \theta) = (r\cos\theta, r\sin\theta)$ (Fig. **5.17**).

Solution   5.3.   (i) $\iint_D 2 - 3x + xy \, dy \, dx = \int_{0}^{1} \int_{0}^{3x} 2 - 3x + xy \, dy \, dx = \frac{9}{8}$. (ii) $\iint_D 2 - 3x + xy \, dx \, dy = \int_{0}^{3} \int_{\frac{y}{3}}^{1} 2 - 3x + xy \, dx \, dy = \frac{9}{8}$. (iii) Step 1: Rotation of 180 degrees counter-clockwise; Step 2: Rotation of 90 degrees clockwise. (Fig. **5.18**).

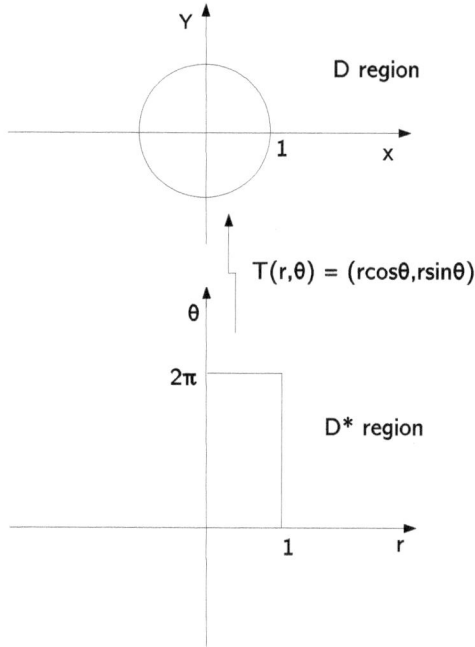

Fig. (5.17). The T mapping from D^* space to D space.

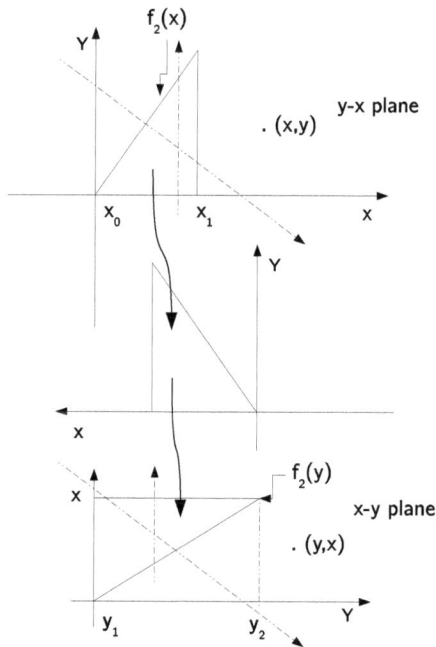

Fig. (5.18). Steps to transform the yx-plane into the xy-plane.

Solution 5.4. (i) $\iiint_D dz\,dy\,dx = \int_{-1}^1 \int_{-\sqrt{1-x^2}}^{\sqrt{1-x^2}} \int_{x^2+y^2}^{\sqrt{x^2+y^2}} dz\,dy\,dx = \int_{-1}^1 \int_{-\sqrt{1-x^2}}^{\sqrt{1-x^2}}$

$\sqrt{x^2+y^2} - x^2 + y^2 \, dy \, dx = \int_{-1}^1 \dfrac{3x^2 \, arsinh\left(\frac{\sqrt{1-x^2}}{|x|}\right)+(1-4x^2)\sqrt{1-x^2}}{3} \, dx =$

$$\left[-\frac{x\left(-2x^2|x|arsinh\left(\frac{\sqrt{1-x^2}}{|x|}\right) - \arcsin(\sqrt{1-x^2}) + \sqrt{1-x^2}(2x^2-3)|x|\right)}{6|x|} \right]_{-1}^{1}$$

$= \frac{\pi}{6}$. (ii) Let $T(r,\theta,z) = (r\cos\theta, r\sin\theta, z)$ $(r,\theta,z) \subset [0,1] \times [0,2\pi] \times [r^2, r]$.

(iii) $J(T(r,\theta,z) = r;$ $\int_0^1 \int_0^{2\pi} \int_{r^2}^{r} f \circ T(r,\theta,z) \, |J(T(r,\theta,z)| dr \, d\theta \, dz =$

$\int_0^1 \int_0^{2\pi} \int_{r^2}^{r} r \, dr \, d\theta \, dz = \frac{\pi}{6}$. (iv) The closed region $[0,1] \times [0,2\pi] \times [r, r^2]$ is transformed into the close D-region by $T(r,\theta,z) = (r\cos\theta, r\sin\theta, z)$ (Fig. **5.19**).

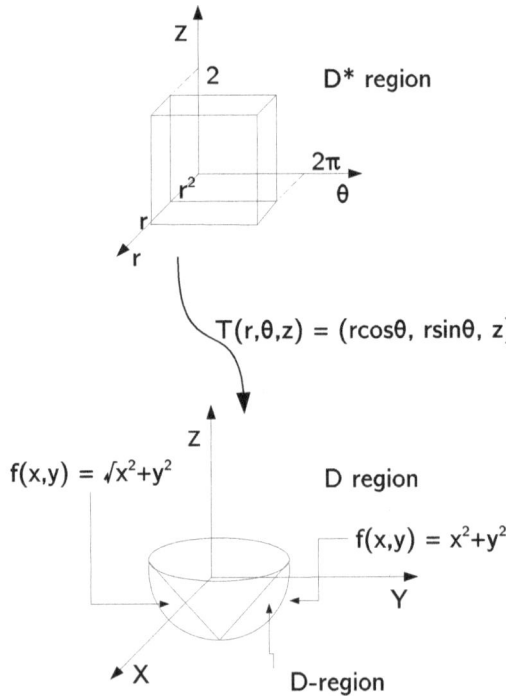

Fig. (5.19). The T mapping from D^* space to D space.

Solution 5.5. (i) $\iiint_D dz\, dy\, dx = \int_{-1}^{1} \int_{-\sqrt{1-x^2}}^{\sqrt{1-x^2}} \int_0^2 dz\, dy\, dx = 2\pi$. (ii) Let
$T(r,\theta,z) = (r\cos\theta, r\sin\theta, z)\ (r,\theta,z) \subset [-1,1] \times [0,2\pi] \times [0,2]$.
(iii) $J(T(r,\theta,z) = r;\ \int_0^1 \int_0^{2\pi} \int_0^2 f \circ T(r,\theta,z)\ |J(T(r,\theta,z)|dr\, d\theta\, dz =$
$\int_0^2 \int_0^{2\pi} \int_0^1 r\ dr d\theta\, dz = 2\pi$. (iv) Fig. (**5.20**).

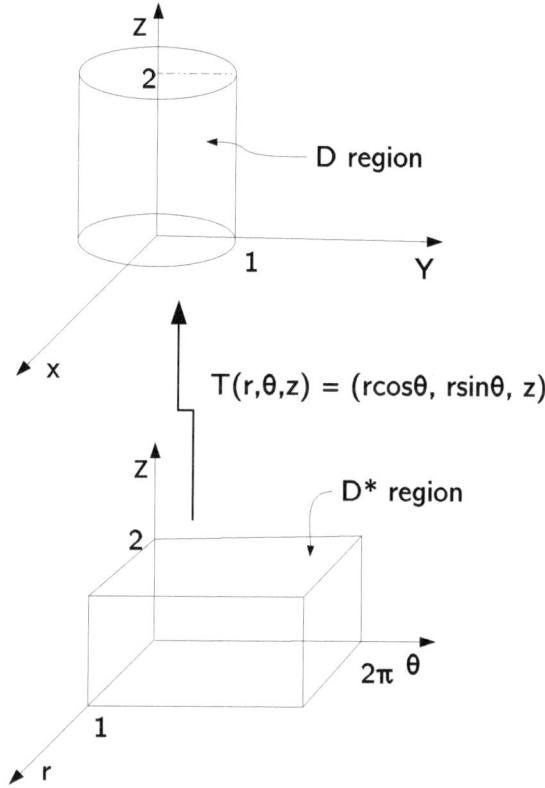

Fig. (5.20). The T mapping from D^* space to D space.

SOLUTIONS FOR CHAPTER 6

Solution 6.1. (i) $\int_0^1 \int_0^1 \frac{1}{xy} dy\, dx = \lim_{(\alpha,\beta)\to(0,0)} \int_{a+\alpha}^{b-\alpha} \int_{\phi_1(x)+\beta}^{\phi_2(x)-\beta} \frac{1}{xy} dy\, dx$
$\lim_{(\alpha,\beta)\to(0,0)} \int_\alpha^{1-\alpha} \int_\beta^{1-\beta} \frac{1}{xy} dy\, dx = \lim_{(\alpha,\beta)\to(0,0)} (\ln(1-\beta) - \ln\beta)(\ln(1-\alpha) - \ln\alpha)$.
(ii) The limit does not exist, therefore, f is not integrable. (iii) Fig. (**6.4**).

Solution 6.2. (i) $\int_1^\infty \int_1^\infty \frac{1}{xy} \, dy \, dx = \lim_{(\alpha,\beta)\to(0,0)} \int_{a+\alpha}^{b-\alpha} \int_{\phi_1(x)+\beta}^{\phi_2(x)-\beta} \frac{1}{xy} \, dy \, dx$

$\lim_{(\alpha,\beta)\to(0,0)} \int_1^\alpha \int_1^\beta \frac{1}{xy} \, dy \, dx = \lim_{(\alpha,\beta)\to(\infty,\infty)} (\ln\beta - \ln 1)(\ln\alpha - \ln 1)$. (ii) The limit does not exist, therefore, f is not integrable. (iii) Fig. (**6.5**).

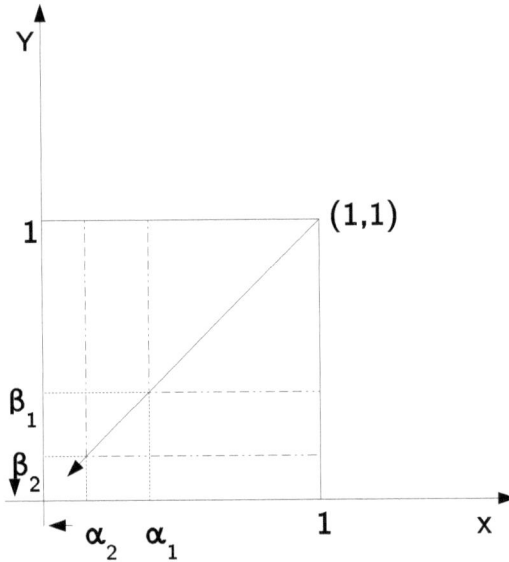

Fig. (6.4). The indeterminations are the lines $x, 0)$ and $(0, y)$. The approximation to the unit square is done through squares that start at the upper right corner.

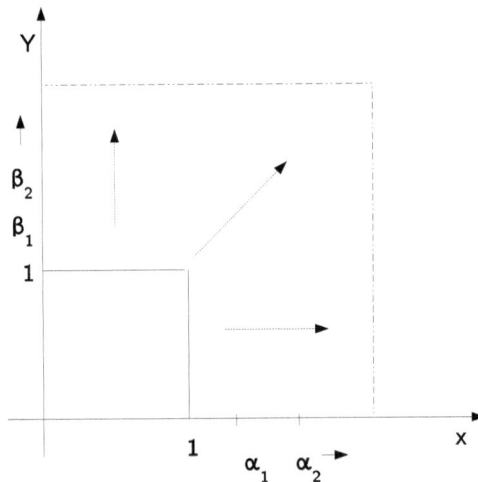

Fig. (6.5). The integration region is Quadrant I, outside the unitary square.

Solution 6.3. (i) $\int_0^1 \int_0^1 \frac{1}{x-y}\, dy\, dx = \lim_{(\alpha,\beta)\to(0,0)} \int_{a+\alpha}^{b-\alpha} \int_{\phi_1(x)+\beta}^{\phi_2(x)-\beta} \frac{1}{x-y}\, dy\, dx$

$\lim_{(\alpha,\beta)\to(0,0)} \int_\alpha^{1-\alpha} \int_\beta^{1-\beta} \frac{1}{x-y}\, dy\, dx = \lim_{(\alpha,\beta)\to(0,0)} \int_\alpha^{1-\alpha} \ln(x-\beta) - \ln(x +$

$\beta - 1)\, dx = \lim_{(\alpha,\beta)\to(0,0)} -(\alpha+\beta)\ln(\alpha+\beta) + (\alpha+\beta)\ln(\alpha+\beta-1) +$

$(\alpha+\beta-1)$. (ii) The limit does not exist, therefore, f is not integrable.
(iii) Fig. (**6.4**).

Solution 6.4. (i) $\int_1^2 \int_1^2 \frac{1}{xy}\, dy\, dx = \lim_{(\alpha,\beta)\to(0,0)} \int_{a+\alpha}^{b-\alpha} \int_{\phi_1(x)+\beta}^{\phi_2(x)-\beta} \frac{1}{xy}\, dy\, dx$

$\lim_{(\alpha,\beta)\to(0,0)} \int_{1+\alpha}^{2-\alpha} \int_{1+\beta}^{2-\beta} \frac{1}{xy}\, dy\, dx = \lim_{(\alpha,\beta)\to(0,0)} (\ln(2-\beta) - \ln(1 +$

$\beta))(\ln(2-\alpha) - \ln(1+\alpha)) = (\ln 2 - \ln 1)(\ln 2 - \ln 1) = 0.4804$. (ii) The
Riemann integral can be treated as an improper integral, but not the other way
round. (iii) Fig. (**6.6**).

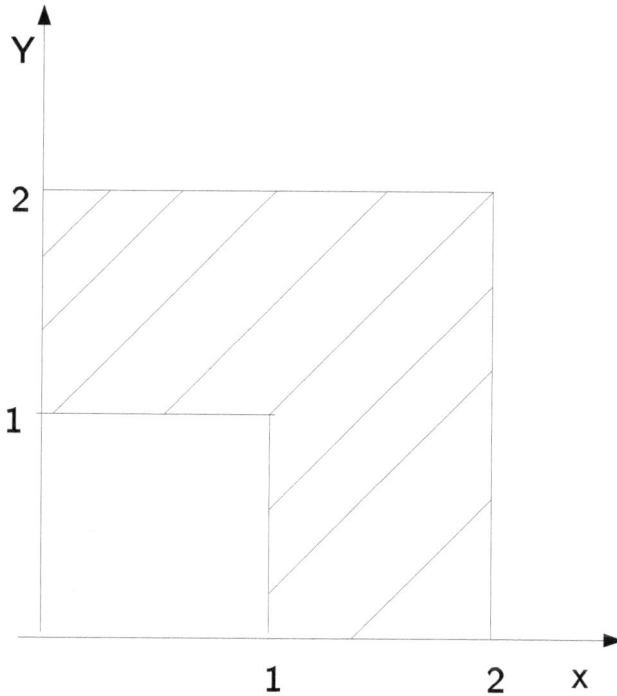

Fig. (6.6). There are no indeterminations.

Solution 6.5. "Convergence of an improper integral depends on how rapidly the
function $f(x)$ tends to zero as $x \to \infty$ (or $x \to -\infty$). Our calculations show that x^{-3}
decreases rapidly enough for convergence, whereas x^{-1} does not" [48].

SOLUTIONS FOR CHAPTER 7

Solution 7.1. (i) The map is $T: \mathbb{R} \Rightarrow \mathbb{R}^2, (a\cos\theta, b\sin\theta), \theta \in [0, 2\pi]$. (ii) See Fig. (7.5).

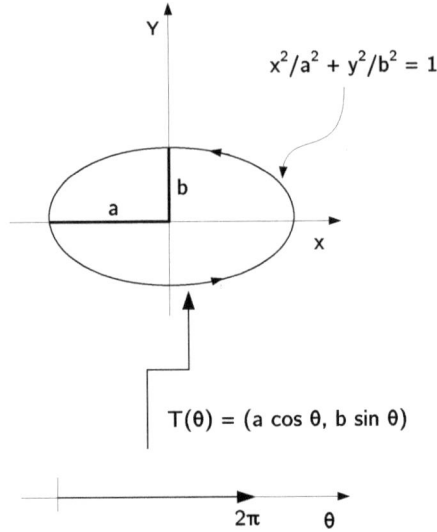

$$x^2/a^2 + y^2/b^2 = 1$$

$$T(\theta) = (a \cos \theta, b \sin \theta)$$

Fig. (7.5). Map of the ellipse where $b < a$.

Solution 7.2. (i) The map is $T: \mathbb{R} \Rightarrow \mathbb{R}^3, (\cos\theta, \sin\theta, 1 - \cos\theta - \sin\theta), \theta \in [0, 2\pi]$. (ii) See Fig. (7.6).

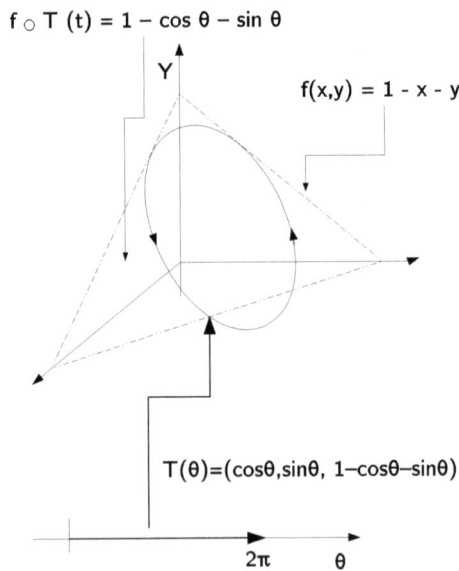

$$f \circ T(t) = 1 - \cos\theta - \sin\theta$$

$$f(x,y) = 1 - x - y$$

$$T(\theta) = (\cos\theta, \sin\theta, 1 - \cos\theta - \sin\theta)$$

Fig. (7.6). Maps of unit circle on the plane.

Solution 7.3. (i) The upper half of the unit sphere is the graph of the function $f(x,y) = \sqrt{1 - x^2 - y^2}$, the map $T: \mathbb{R}^2 \Rightarrow \mathbb{R}^2, (r\cos\theta, r\sin\theta), r \in [0,1], \theta \in [0,2\pi]$ transforms the rectangle into the unit circle, and $f \circ T$ is the third component of the map $T: \mathbb{R}^2 \Rightarrow \mathbb{R}^3$. Then,

$$T(r,\theta) = (r\cos\theta, r\sin\theta, \sqrt{1 - r^2\cos^2\theta - r^2\sin^2\theta}) = (r\cos\theta, r\sin\theta, \sqrt{1 - r^2}).$$

(ii) See Fig. **(7.7)**.

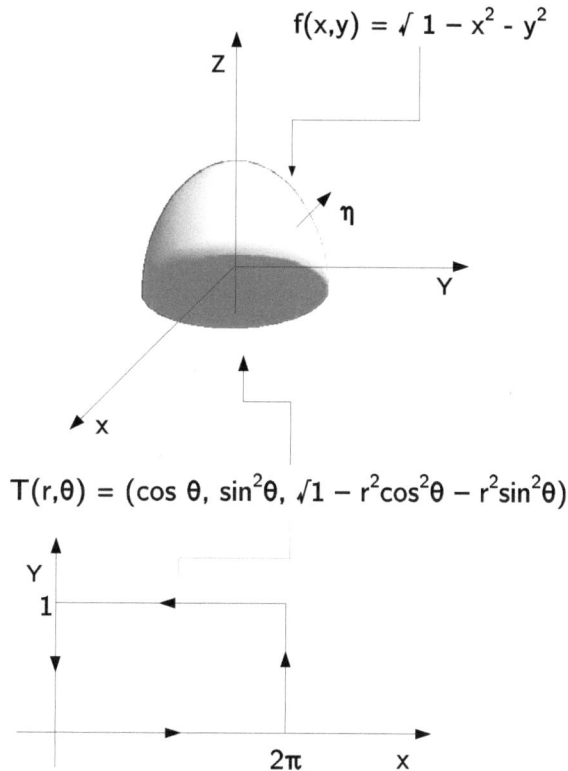

Fig. (7.7). Map of the unit circle in the plane.

Solution 7.4. (i) $\oint_C F \circ c(t) \cdot c'(t)\, dt = \int_0^{2\pi} (-\sin t, \cos t) \cdot (-\sin t, \cos t) dt = \int_0^{2\pi} dt = 2\pi$. (ii) $\oint_C F \circ c(t) \cdot c'(t)\, dt = \int_{2\pi}^0 (-\sin t, \cos t) \cdot (-\sin t, \cos t) dt = \int_{2\pi}^0 dt = -2\pi$.

Solution 7.5. (i) $\oint_C F \circ c(t) \cdot c'(t)\, dt = \int_0^{2\pi} (-\sin t, \cos t, 1) \cdot (-\sin t, \cos t, 1) dt = \int_0^{2\pi} dt = 4\pi$.

(ii) $\oint_C F \circ c(t) \cdot c'(t)\, dt = \int_{2\pi}^0 (-\sin t, \cos t, 1) \cdot (-\sin t, \cos t, 1)dt = \int_{2\pi}^0 dt = -2\pi$.

Solution 7.6. (i) Let $T(r,\theta) = (r\cos\theta, r\sin\theta, \sqrt{1-r^2})$ (Solution 0). Then $_sF \circ T(u,v) \cdot \eta(u,v)\, dS = \int_0^{2\pi} \int_0^1 (r\cos\theta, r\sin\theta, \sqrt{1-r^2}) \cdot (\frac{-r^2\cos\theta}{\sqrt{1-r^2}}, \frac{r^2\sin\theta}{\sqrt{1-r^2}}, r)\, dr\, d\theta = \int_0^{2\pi} \int_0^1 \frac{r}{\sqrt{1-r^2}} dr\, d\theta = -2\pi[\sqrt{1-x^2}]_0^1 = 2\pi$.

(ii) $_sF \circ T(u,v) \cdot \eta(u,v)\, dS = \int_0^{2\pi} \int_0^1 (r\cos\theta, r\sin\theta, \sqrt{1-r^2}) \cdot -(\frac{-r^2\cos\theta}{\sqrt{1-r^2}}, \frac{r^2\sin\theta}{\sqrt{1-r^2}}, r)\, dr\, d\theta = -\int_0^{2\pi} \int_0^1 \frac{r}{\sqrt{1-r^2}} dr\, d\theta = -2\pi[\sqrt{1-x^2}]_0^1 = -2\pi$.

Solution 7.7. (i) Let $c(t) = (t, t^2)$, $\int_C f \circ c(t) \|c'(t)\|\, dt = \int_a^b f(c(t)) \|c'(t)\|\, dt = \int_0^{2\pi}$

$$(t^2 + t^4)\sqrt{1 + 4t^2}\, dt = \frac{\sqrt{4x^2 + 1}(256x^5 + 400x^3 + 42x) - 21\sinh 21}{1536}.$$

Solution 7.8. $\int_0^1 \int_0^1 \sqrt{(1,0,3v) \times (0,1,3u)}\, dv\, du = \int_0^1 \int_0^1 \sqrt{9v^2 + 9u^2 + 1}\, dv\, du = 2.5376$.

SOLUTIONS FOR CHAPTER 8

Solution 8.1. From Green's theorem

$$\oint_{\partial C} F \circ c(t) \cdot c'(t)\, ds = \iint_C (\nabla \times F) \cdot k\, dy\, dx.$$

Then $F(x,y) = (2y + \sqrt{1 + x^5}, 5x - e^{y^2})$, $\nabla \times F = (0,0,3)$. The map T from $\mathbb{R}^2$ to $\mathbb{R}^2$. So $\iint_C (\nabla \times F) \cdot k\, dy\, dx = \int_{-2}^2 \int_0^{\sqrt{1-x^2}} (0,0,3) \cdot (0,0,1)\, dy\, dx = 3 \int_0^2 \int_0^\pi r\, dr\, d\theta = 6\pi$.

Solution 8.2. From Green's theorem

$$\oint_{\partial C} F \circ c(t) \cdot c'(t)\, ds = \iint_C (\nabla \times F) \cdot k\, dy\, dx.$$

The vector field in the integral above is $F(x,y) = (y^2, 3xy)$. We could compute the line integral directly (see below). But, we can compute this integral more easily using Green's theorem to convert the line integral into a double integral. The integrand of the double integral would be (Eq. 8.1).

$$\frac{\partial F_2}{\partial x} - \frac{\partial F_1}{\partial y} = 3y - 2y = y. \tag{8.1}$$

Since the line integral was over the boundary of the half disc, the region of integration for the double integral is the half-disc D itself. (Since C was oriented counter-clockwise, the orientation matches; otherwise, we would have had to multiply by negative one to get the correct sign). The region D is described by Fig. (**8.6**), where $-1 < x < 1$ and $0 < y < \sqrt{1 - x^2}$.

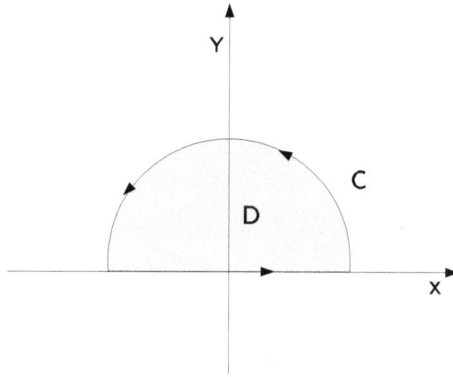

Fig. (8.6). Upper-half of the unit disc D.

$$\oint_C y^2 dx + 3xy\, dy = \iint_D \left(\frac{\partial F_2}{\partial x} - \frac{\partial F_1}{\partial y}\right) dA$$

$$= \iint y\, dA$$

$$= \int_{-1}^{1} \int_0^{\sqrt{1-x^2}} y\, dy\, dx$$

$$= \int_{-1}^{1} \left(\frac{y^2}{2}\Big|_{y=0}^{y=\sqrt{1-x^2}}\right) dx$$

$$= \int_{-1}^{1} \frac{1-x^2}{2}\, dx$$

$$= \frac{x}{2} - \frac{x^3}{6}\Big|_{-1}^{1} = \frac{2}{3}.$$

Alternative Solution method: You could also compute this line integral directly without using Green's theorem to get the same answer. However, in this case, the integral is more difficult. We have to compute the integral in two parts: the first part is the half circle, oriented from right to left (labelled C_1) and the second part is the line segment, oriented from left to right (labelled C_2) (Fig. **8.7**).

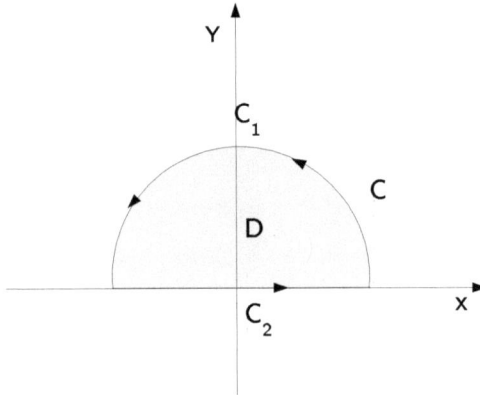

Fig. (8.7). Upper-half of the unit disk D.

First, calculate the integral alone C_1. Parameterise C_1 by $c(t) = (\cos t, \sin t)$, $0 < x < \pi$. Then $c'(t) = (-\sin t, \cos t)$.

Calculating:

$$\int_{C_1} F ds = \int_0^\pi F(c(t)c'(t)\, dt$$

$$= \int_0^\pi F(\cos t, \sin t) \cdot (-\sin t, \cos t) dt$$

$$= \int_0^\pi (\sin^2 t, 3\cos t \sin t) \cdot (-\sin t, \cos t) dt$$

$$= \int_0^\pi (-\sin^3 t + 3\sin t \cos^2 t) dt$$

$$= \int_0^\pi (-\sin t(1 - \cos^2 t) + 3\sin t \cos^2 t) dt$$

$$= \int_0^\pi (-\sin t + 4\sin t \cos^2 t) dt$$

We can calculate that $\int_0^\pi \sin t\, dt = 2$ and (let $u = \cos t$, $du = -\sin t\, dt$).

$$\int_0^\pi \sin t\cos^2 t\, dt = \int_1^{-1} -u^2\, du$$

$$= -\frac{u^3}{3}\Big|_1^{-1} = -(-1)/3 + 1/3 = 2/3.$$

Therefore $\int_{C_1} F ds = -2 + 4(2/3) = -2 + 8/3 = 2/3$. The integral along C_2 is easy. Along C_2, $y = 0$, so that $\int_{C_2} F dx = 0$. Putting this all together, we verify that $\int_C = \int_{C_1} F ds + \int_{C_2} F ds = \frac{2}{3} + 0 = \frac{2}{3}$. Our direct calculation of the line integral agrees with the above result that we obtained by applying Green's theorem to convert the line integral to a double integral.

Solution 8.3. From Stokes' theorem

$$\oint_{\partial D} F \circ c(t) \cdot c'(t)\, ds = \iint_S (\nabla \times F) \cdot T_v \times T_u\, dv\, du.$$

(i) The map T from $\mathbb{R}$ to $\mathbb{R}^3$ is $T(\theta) = (\cos\theta, \sin\theta, -\cos\theta - \sin\theta)$, where $0 < \theta < 2\pi$. Then $\oint_{\partial S} F \circ T(t) \cdot T'(t)\, ds = \int_0^{2\pi} (\sin\theta, -\cos\theta - \sin\theta, \cos\theta) \cdot (-\sin\theta, \cos\theta, \sin\theta - \cos\theta)\, d\theta = \int_0^{2\pi} -\sin^2\theta - \cos^2\theta - \cos\theta\sin\theta + \cos\theta\sin\theta - \cos^2\theta\, d\theta = -3\pi$.

(ii) See Fig. (**8.8**).

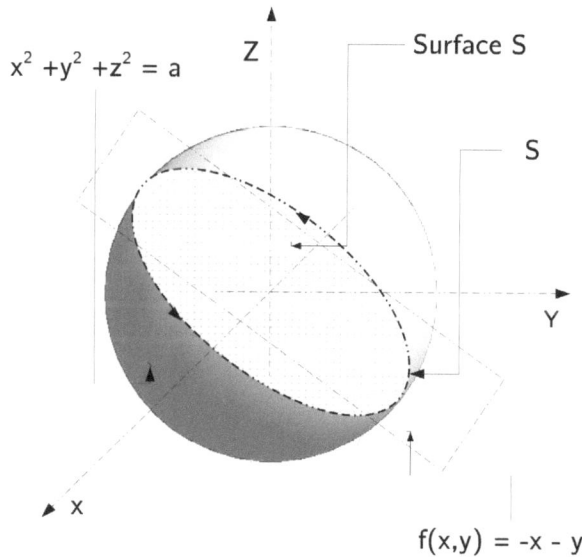

Fig. (8.8). Geometrical description of surfaces S and ∂S.

Solution 8.4. From Stokes' theorem

$$\oint_{\partial D} F \circ c(t) \cdot c'(t) \ ds = \iint_S (\nabla \times F) \cdot T_v \times T_u \ dv \ du.$$

The Maxwell-Faraday equation is one of the four Maxwell's equations. It states that a time-varying magnetic field B will always accompany a spatially varying electric field E [62], *i.e.*

$$\nabla \times E = -\frac{\partial B}{\partial t}.$$

This equation plays a fundamental role in the theory of classical electromagnetism (Fig. **8.9**). It can be written in an integral form (Eq. 8.2)

$$\oint_{\partial L} E \ dL = \iint_A \nabla \times E \ dA = \frac{\partial}{\partial t} \iint_A B \ dA. \tag{8.2}$$

where

$$\oint_L E \ dL = \text{voltage around } L,$$

and

$$\iint_A B \ dA = \text{magnetic fluxa cross } A.$$

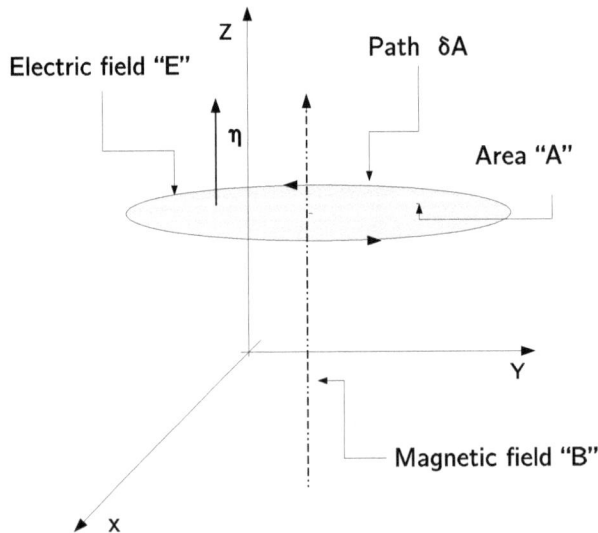

Fig. (8.9). Geometric interpretation of the electric field and the magnetic field.

Solution 8.5. From Gauss' theorem

$$\oiint_{\partial W} F \circ T(u,v) \cdot T_v \times T_u \, dv \, du = \iiint_W (\nabla \cdot F) \, dz \, dy \, dx.$$

Then $\quad\oiint_{\text{surfacesphere}} 3xi + 2yj \, dx \, dy = \iiint_{\text{volumesphere}} 5 \, dz \, dy \, dx = 5[\frac{4}{3}\pi(4^3)] \approx 426\pi.$

Solution 8.6. From Gauss' theorem

$$\oiint_{\partial W} F \circ T(u,v) \cdot T_v \times T_u \, dv \, du = \iiint_W (\nabla \cdot F) \, dz \, dy \, dx.$$

Note that the surface integral will be difficult to compute, since there are six different components to parameterise (corresponding to the six sides of the box), and we would have to compute six different integrals. Instead, using Gauss' theorem, it will be easier to compute the integral of $\nabla \cdot F$ over B. First we compute $\nabla \cdot F = 2xz^3 + 2xz^3 + 4xz^3 = 8xz^3$. Now we integrate this function over the region B bounded by S:

$$\oiint_s F \, ds = \iiint_B (\nabla \cdot F) dV \qquad (8.3)$$

$$= \int_{-3}^{3} \int_{-2}^{2} \int_{-1}^{1} 8xz^3 dx \, dy \, dz$$

$$= 0.$$

SOLUTIONS FOR CHAPTER 9

Solution 9.1. (i)

$$\begin{aligned} w \wedge \eta &= (3dx + dy)(e^x dx + 2dy) \\ &= 3e^x dx \wedge dx + 6dx \wedge dy + e^x dy \wedge dx + 2dy \wedge dy \\ &= (6 - e^x)dx \wedge dy \\ &= (6 - e^x)dxdy \end{aligned} \qquad (9.4)$$

(ii)

$$\begin{aligned} d(6 - e^x) \wedge dx \wedge dy &= -e^x dx \wedge dx \wedge dy \\ &= 0 \end{aligned} \qquad (9.5)$$

Solution 9.2.

$$
\begin{aligned}
dw &= d(x_1 + x_3^2) \wedge dx_1 \wedge dx_2) \\
&= dx_1 dx_1 dx_2 + 2x_3 dx_1 dx_2 \\
&= 2x_3 dx_1 dx_2 dx_3 \\
&= 2x_3 dx_1 \wedge dx_2 \wedge dx_3
\end{aligned}
$$ (9.6)

Solution 9.3.

$$
\begin{aligned}
dx \wedge dy &= (-r\sin\theta d\theta + \cos\theta dr) \wedge (r\cos\theta d\theta + \sin\theta dr) \\
&= -r^2\sin\theta\cos\theta d\theta d\theta - r\sin^2\theta d\theta dr + r\cos^2\theta dr d\theta + \cos\theta\sin\theta dr dr \\
&= (-r\sin^2\theta - r\cos^2\theta)d\theta \wedge dr \\
&= r\, dr \wedge d\theta
\end{aligned}
$$ (9.7)

Solution 9.4.

(i)
$$
\begin{aligned}
dw &= d(xdx + yzdy + x^2ydz) \\
&= d(xdx) + d(yzdy) + d(x^2ydz) \\
&= d(x)dx + d(yz)dy + d(x^2y)dz \\
&= dxdx + zdydy + ydzdy + 2xydxdz + x^2dydz \\
&= ydzdy + 2xydxdz + x^2dydz \\
&= -ydydz + 2xydxdz + x^2dydz \\
&= 2xydxdz + (x^2 - y)dydz \\
&= 2xydx \wedge dz + (x^2 - y)dy \wedge dz
\end{aligned}
$$ (9.8)

$$
\begin{aligned}
d(xdx) &= (x)'_x dx + (x)'_y dx + (x)'_z dx \\
&= dxdx + 0 + 0 \\
&= dxdx
\end{aligned}
$$ (9.9)

$$
\begin{aligned}
d(yzdy) &= (yz)'_x dy + (yz)'_y dy + (yz)'_z dy \\
&= 0 + zdydy + ydzdy
\end{aligned}
$$ (9.10)

$$
\begin{aligned}
d(yzdy) &= (x^2y)'_x dz + (x^2y)'_y dz + (x^2y)'_z dz \\
&= 2xydxdz + x^2dydz + 0
\end{aligned}
$$ (9.11)

(ii)

$$
\begin{aligned}
d(dw) &= d[2xydxdz + (x^2 - y)dydz] \\
&= d(2xydxdz) + d[(x^2 - y)dydz] \\
&= d(2xy)dxdz + d(x^2 - y)dydz \\
&= 2ydxdxdx + 2xdydxdz + 2xdxdydz - dydydz \qquad \textbf{(9.12)} \\
&= 2xdydxdz + 2xdxdydz \\
&= -2xdxdydz + 2xdxdydz \\
&= 0
\end{aligned}
$$

$$
\begin{aligned}
d(2xy)dxdz &= (2xy)'_x dxdz + (2xy)'_y dxdz + (2xy)'_z dxdz \\
&= 2ydxdxdx + 2xdydxdz + 0 \qquad\qquad \textbf{(9.13)}
\end{aligned}
$$

$$
\begin{aligned}
d(x^2 - y)dydz &= (x^2 - y)'_x dydz + (x^2 - y)'_y dydz + (x^2 - y)'_z dydz \\
&= 2xdxdydz - dydydz + 0 \qquad\qquad \textbf{(9.14)}
\end{aligned}
$$

(iii)

$$
\begin{aligned}
(w \wedge \eta) &= (xdx + yzdy + x^2ydz) \wedge (xydz) \\
&= x^2ydxdz + y^2zxdydz + x^3yx^2dzdz \qquad \textbf{(9.15)} \\
&= x^2ydxdz + y^2zxdydz
\end{aligned}
$$

(iv)

$$
\begin{aligned}
d(w \wedge \eta) &= d(x^2ydxdz + y^2zxdydz) \\
&= d(x^2ydxdz) + d(y^2zxdydz) \\
&= d(x^2y)dxdz + d(y^2zx)dydz \qquad \textbf{(9.16)} \\
&= -x^2dxdydz + y^2zdxdydz \\
&= (y^2z - x^2)dxdydz
\end{aligned}
$$

$$
\begin{aligned}
d(x^2y)dxdz &= (x^2y)'_x dxdz + (x^2y)'_y dxdz + (x^2y)'_z dxdz \\
&= 2xydxdxdz + x^2dydxdz + 0 \qquad\qquad \textbf{(9.17)} \\
&= -x^2dxdydz
\end{aligned}
$$

$$
\begin{aligned}
d(y^2 zx)dydz &= (y^2 zx)'_x dydz + (y^2 zx)'_y dydz + (y^2 zx)'_z dydz \\
&= y^2 z dxdydz + 2yxz dydydz + y^2 x dzdydz \\
&= y^2 z dxdydz
\end{aligned}
\tag{9.18}
$$

(v)

From (iv), $d(w \wedge \eta) = dw \wedge \eta + (-1)^k w \wedge d\eta$.

SOLUTIONS FOR CHAPTER 10

Solution 10.1.

$$
\begin{aligned}
\int_c \frac{-y}{x^2+y^2}dx + \frac{x}{x^2+y^2}dy &= \int_0^{2\pi} -\sin\theta\cos'\theta d\theta + \cos\theta\sin'\theta d\theta \\
&= \int_0^{2\pi} d\theta \\
&= 2\pi
\end{aligned}
\tag{10.19}
$$

Solution 10.2. (i) $P(x,y,z) = 0$, $Q(x,y,z) = 0$, and $R(x,y,z) = x$. (ii) $x(r,\theta) = r\cos\theta$, $y(r,\theta) = r\sin\theta$, and $z(r,\theta) = 0$. (iii) $P \circ T(r,\theta) = 0$, $Q \circ y(r,\theta) = 0$, and $R \circ z(r,\theta) = r\cos\theta$. (iv) $\frac{\partial(x,y)}{\partial(r,\theta)} = r$, $\frac{\partial(y,z)}{\partial(r,\theta)} = 0$, and $\frac{\partial(z,x)}{\partial(r,\theta)} = 0$. (v) $\int_{\theta_0}^{\theta_1} \int_{r_0}^{r_1} P \circ T(r,\theta) \frac{\partial(y,z)}{\partial(\theta,r)} + Q \circ T(r,\theta) \frac{\partial(z,x)}{\partial(r,\theta)} + R \circ T(\theta,r) \frac{\partial(x,y)}{\partial(\theta,r)} dr \, d\theta = \int_0^{2\pi} \int_0^1 r^2 \cos\theta \, dr \, d\theta = 0.$

Note 10.1 $\frac{\partial(x,y)}{\partial(r,\theta)} = \begin{vmatrix} \frac{\partial x}{\partial r} & \frac{\partial x}{\partial \theta} \\ \frac{\partial y}{\partial r} & \frac{\partial y}{\partial \theta} \end{vmatrix}$

Solution 10.3.

$$
\begin{aligned}
\iint_D \left(\frac{\partial Q}{\partial x} - \frac{\partial P}{\partial y}\right) dydx &= \int_0^1 \int_0^1 2x - 1 dy \, dx \\
&= \int_0^1 2x - 1 dx \\
&= 0
\end{aligned}
$$

$$
\begin{aligned}
\int_{\partial D_1} P \circ c_1(t) \frac{x(t)}{dt} + Q \circ c_1(t) \frac{y(t)}{dt} & \\
+ \int_{\partial D_2} P \cdot c_2(t)\frac{dx}{dt} + Q \cdot c_2(t)\frac{dy}{dt} & \\
+ \int_{\partial D_3} P \cdot c_3(t)\frac{dx}{dt} + Q \cdot c_3(t)\frac{dy}{dt} & \\
+ \int_{\partial D_4} P \cdot c_4(t)\frac{dx}{dt} + Q \cdot c_4(t)\frac{dy}{dt} & \\
= 0 + 1 - 1 + 0 & \\
= 0 &
\end{aligned}
\tag{10.20}
$$

where $c_1(t) = (t, 0), t \in [0,1]$, $c_2(t) = (1, t), t \in [0,1]$, $c_3(t) = (t, 1), t \in [1,0]$ and $c_4(t) = (0, t), t \in [1,0]$.

$$\int_{\partial D_1} P \cdot c_1(t) \frac{dx}{dt} + Q \cdot c_1(t) \frac{dy}{dt} = \int_0^1 xy(0) + x^2 y^2(1) dt$$
$$= \int_0^1 0(1) + 0(t^2) dt$$
$$= 0$$

$$\int_{\partial D_2} P \cdot c_2(t) \frac{dx}{dt} + Q \cdot c_2(t) \frac{dy}{dt} = \int_0^1 0(t) + 1(1) dt$$
$$= 1 \tag{10.21}$$

$$\int_{\partial D_3} P \cdot c_3(t) \frac{dx}{dt} + Q \cdot c_3(t) \frac{dy}{dt} = \int_1^0 1(1) + 0(t^2) dt$$
$$= -1$$
$$\int_{\partial D_4} P \cdot c_4(t) \frac{dx}{dt} + Q \cdot c_4(t) \frac{dy}{dt} = \int_1^0 0(1) + 0(1) dt$$
$$= 0$$

Solution 10.4. If $T(\theta, r) = (r\cos t, r\sin t, 1 - r\cos t - r\sin t)$,

$$\frac{\partial(T_y, T_z)}{\partial(\theta, r)} = \begin{vmatrix} r\cos\theta & \sin\theta \\ r\sin\theta & -r\cos\theta \end{vmatrix} \tag{10.22}$$

$$\frac{\partial(T_z, T_x)}{\partial(\theta, r)} = \begin{vmatrix} r\sin\theta & -r\cos\theta \\ -r\sin\theta & \sin\theta \end{vmatrix} \tag{10.23}$$

$$\frac{\partial(T_x, T_y)}{\partial(\theta, r)} = \begin{vmatrix} -r\sin\theta & \sin\theta \\ r\cos\theta & \sin\theta \end{vmatrix} \tag{10.24}$$

$$\iint_S dw = \int_0^1 \int_0^{2\pi} \left(\frac{\partial R}{\partial y} - \frac{\partial Q}{\partial z}\right) dy dz + \left(\frac{\partial P}{\partial z} - \frac{\partial R}{\partial x}\right) dz dx + \left(\frac{\partial Q}{\partial x} - \frac{\partial P}{\partial y}\right) dx dy$$
$$= \int_0^1 \int_0^{2\pi} \left(\frac{\partial R}{\partial y} - \frac{\partial Q}{\partial z}\right) \circ T(\theta, r) \frac{\partial(T_y, T_z)}{\partial(\theta, r)}$$
$$+ \left(\frac{\partial P}{\partial z} - \frac{\partial R}{\partial x}\right) \circ T(\theta, r) \frac{\partial(T_z, T_x)}{\partial(\theta, r)}$$
$$+ \left(\frac{\partial Q}{\partial x} - \frac{\partial P}{\partial y}\right) \circ T(\theta, r) \frac{\partial(T_x, T_y)}{\partial(\theta, r)} \tag{10.25}$$
$$= \int_0^1 \int_0^{2\pi} (0) \frac{\partial(T_y, T_z)}{\partial(\theta, r)} + (0) \frac{\partial(T_z, T_x)}{\partial(\theta, r)} + (0) \frac{\partial(T_x, T_y)}{\partial(\theta, r)} d\theta dr$$
$$= 0$$

$$\int_{\partial D} P \cdot c(t) \frac{dx}{dt} + Q \cdot c(t) \frac{dy}{dt} + R \cdot c(t) \frac{dy}{dt} \quad = \int_0^{2\pi} \; (\cos t)(-\sin t)$$
$$+ (\sin t)(\cos t)$$
$$+ (1 - \cos t - \sin t)(\sin t - \cos t)dt \tag{10.26}$$
$$= 0$$

Solution 10.5.

$$\int_{\Omega} \quad = \iiint_{\Omega} \left(\frac{\partial P}{\partial x} + \frac{\partial Q}{\partial y} + \frac{\partial R}{\partial z} \right) dz\, dy\, dx$$
$$= \int_{-1}^{-1} \int_{-\sqrt{1-x^2}}^{\sqrt{1-x^2}} \int_{-\sqrt{1-x^2-y^2}}^{\sqrt{1-x^2-y^2}} 8xyz\, dz\, dy\, dx \tag{10.27}$$
$$= 0$$

$$\iint_{\partial \Omega} d\Omega \quad = \int_0^{\pi} \int_0^{2\pi} P \circ T \frac{\partial(T_y, T_z)}{\partial(\theta, \phi)} + Q \circ T \frac{\partial(T_z, T_x)}{\partial(\theta, \phi)} + R \circ T \frac{\partial(T_x, T_y)}{\partial(\theta, \phi)} \, d\theta\, d\phi$$
$$= \int_0^{\pi} \int_0^{2\pi} \cos^2\theta \sin^2\phi \frac{\partial(T_y, T_z)}{\partial(\theta, \phi)} + \sin^2\theta \sin^2\phi \frac{\partial(T_z, T_x)}{\partial(\theta, \phi)} + \cos^2\phi$$
$$\frac{\partial(T_x, T_y)}{\partial(\theta, \phi)} d\theta d\phi \tag{10.28}$$
$$= \int_0^{\pi} \int_0^{2\pi} \; (\cos^2\theta \sin^2\phi)(-\cos\theta \sin^2\phi) + (\sin^2\theta \sin^2\phi)(-\sin^2\phi)$$
$$+ (\cos^2\phi)(-\sin^2\theta \sin\phi\cos\phi - \cos^2\theta \sin\phi\cos\phi)d\theta d\phi$$
$$= 0$$

Note 10.2. The sign depends on the orientation.

$$\frac{\partial(T_y, T_z)}{\partial(\theta, r)} \quad = \begin{vmatrix} \cos\theta\cos\phi & \sin\theta\cos\phi \\ 0 & -\sin\phi \end{vmatrix} \tag{10.29}$$
$$= -\cos\theta \sin^2\phi$$

$$\frac{\partial(T_z, T_x)}{\partial(\theta, r)} \quad = \begin{vmatrix} 0 & -\sin\phi \\ -\sin\theta\sin\phi & \cos\theta\cos\phi \end{vmatrix} \tag{10.30}$$
$$= -\sin^2\phi$$

$$\frac{\partial(T_x, T_y)}{\partial(\theta, r)} \quad = \begin{vmatrix} -\sin\theta\sin\phi & \cos\theta\cos\phi \\ \cos\theta\sin\phi & \sin\theta\cos\phi \end{vmatrix} \tag{10.31}$$
$$= -\sin^2\theta \sin\phi\cos\phi - \cos^2\theta \sin\phi\cos\phi$$

REFERENCES

[1]   P. E. Hand, "Problem on relative motion and vector addition," 2018, http://www.leadinglesson.com/problem-on-relative-motion-and-vector-addition.

[2]   J. V. Becerril, J. Grabinsky, and J. Guzmán, *Problemario de vectores, rectas, planos, sistemas de ecuaciones lineales, cónicas y esferas Con anexo. Problem 37*. Universidad Autónoma Metropolitana, 2004.

[3]   P. E. Hand, "Problem on gradient, directional derivative and level curves," 2018, http://www.leadinglesson.com/problem-on-gradient-directionalderivative-and-level-curves.

[4]   U. de los Andes, "Funciones de varias variables," 2009, http://www.ciens.ula.ve/matematica/publicaciones/guias/servicio docente/maria victoria/2009/texto21/funciones varias variables.pdf.

[5]   P. E. Hand, "Problem on optimization without constraint," 2018, http://www.leadinglesson.com/problem-on-optimization-without-constraint.

[6]   U. de las Palmas, "Soluciones a los ejercicios propuestos: Matemáticas iii. curso 08–09 problema 19," 2009, https://www2.ulpgc.es/hege/almacen/download/7065/7065606/soluciones problemas 08 09 temas 8 al 10.pdf.

[7]   P. E. Hand, "Problem on finding a potential function of a vector field," 2018, http://www.leadinglesson.com/problem-on-finding-a-potentialfunction-of-a-vector-field.

[8]   P. L. Clark, *Handout Five: Vector Fields*. Department of Mathematics, University of Georgia, 2009.

[9]   P. E. Hand, "Problem on a double integral over a circle," 2018, http://www.leadinglesson.com/problem-on-a-double-integral-over-a-circle.

[10]   F. F. Vilches-Medina, "Problemas resueltos y propuestos de cálculo en varias variables." 2011, https://www.ucursos.cl/usuario/b4eb6d37062854338662ba7470704112/mi blog/r/prp cvv.pdf.

[11]   K. J. Mitchell, "Improper integrals with infinite discontinuities," 2013, http://math.hws.edu/mitchell/Math131S13/tufte-latex/Improper2.pdf.

[12]   B. Ikenaga, "Improper integrals," 2018, http://sites.millersville.edu/bikenaga/calculus/improper-integral/improperintegral.html.

[13]   P. E. Hand, "Problem on evaluating a line integral," 2018, http://www.leadinglesson.com/problem-on-evaluating-a-line-integral.

[14]   M. Brittenham, "The surface area of a torus (*i.e*, doughnut)." 2012, https://www.math.unl.edu/m~brittenham2/classwk/208s12/inclass/surface.area. of.a.torus.pdf.

[15]   P. E. Hand, "Problem on a line integral over a circle," 2018, http://www.leadinglesson.com/problem-on-a-line-integral-over-a-circle.

[16]   U. N. de la Plata, "Campos vectoriales - parte b," 2017, http://www.mate.unlp.edu.ar/practicas/54 12 16112017121618.pdf.

[17]   S. Ramos, J. A. Juárez, and G. Sobczyk, "From vectors to geometric algebra 4, page 17." 2018, https://arxiv.org/pdf/1802.08153.pdf.

[18]   Wikipedia, "Triple product," 2018, https://en.wikipedia.org/wiki/Triple product.

[19]   SlideShare, "Triple product," 2018, https://es.slideshare.net/edvinogo/5-tripleproducto-escalar.

[20]   Wikipedia, "Polar coordinate system," 2017, https://en.wikipedia.org/w/index. php?title=Polar coordinate system&oldid=809014788.

[21]   "Cylindrical coordinate system," 2018, https://en.wikipedia.org/w/inde x.php?title=Cylindrical coordinate system&oldid=811970435.

[22]   "Spherical coordinate system," 2018, https://en.wikipedia.org/w/index. php?title=Spherical coordinate system&oldid=817618690.

[23]   "Vector columna," 2018, https://es.wikipedia.org/wiki/Vector columna.

[24]   "Letter frequency," 2018, https://en.wikipedia.org/wiki/Letter frequency.

[25]   J. Marsden and A. Tromba, *Vector Calculus*. New York, NY 10004, USA:WH Freeman And Company, 2011.

[26]   Wikipedia, "Indeterminate form," 2018, https://en.wikipedia.org/wiki/Indeterminate form#Indeterminate form 0/0.

[27]   C. A. D. Prado, "Tema 7, reglas de l'hospital," 2018, http://www.ugr.es/ camilo/calculo-ii-grado-en-matemat/apuntes/tema-7.pdf.

[28]   S. Lang, *Calculus of Several Variables*. New York, NY 10013, USA: Springer Verlag., 1987.

[29]   J. P.Merx, "Math counterexample," 2018, http://www.mathcounterexamples.net/differentiability-multivariable-real-functions-part1/.

[30]   P. Laval, "Mathematics," 2017, http://ksuweb.kennesaw.edu/plaval/math2203/funcnD limcont.pdf.

[31]   M. Corral, "Vector calculus," 2018, http://omega.albany.edu:8008/mat214dir/ CoralCalc3BOOK.pdf.

[32]   A. Dendane, "Freemathematics tutorials, calculus mutivariable, critical points, examples," 2003, http://www.analyzemath.com/calculus/multivariable/critical points.html.

[33]   A. Lugon, "Breve sobre el hessiana orlado," 2018, http://macareo.pucp.edu.pe /alugon/teaching/Optimizacion/hessiano%20orlado.pdf.

[34]   O. de Oliveira, "The implicit and the inverse function theorems: Easy proofs," 2018, https://www.ime.usp.br/ oliveira/IMPLI-1-RAEX-FINAL.pdf.

[35]   J. Llopis, "Cálculo de extremos de funciones de varias variables," 2018, https: //www.matesfacil.com/UNI/varias variables/extremos/extremos-varias-variables.html.

[36]   B. González-Rodriguez, M. Hernández-Abreu, M. Jiménez-Paiz, A. Marrero-Rodríguez, and A. Sanabria-García, *Funciones de varias variables: problemas resueltos*. Santa Cruz de Tenerife, 32800, España: Universidad de la Laguna., 2018, https://campusvirtual.ull.es/ocw/pluginfile.php/6189/mod resource/content/1/tema3/PR3-varvariables.pdf.

[37]   L. University, "Vector fields," 2018, http://tutorial.math.lamar.edu/Classes/CalcIII/VectorFields.aspx.

[38]   K. Kuniyuki, "Math notes and math tests," 2018, http://www.kkuniyuk.com/Math252FlowLines.pdf.

[39]   H. Mathematics, "Curl and divergence," 2018, http://www.math.harvard.edu/archive/21a spring 09/PDF/13-05-curl-anddivergence.pdf.

[40]   *What Is the Meaning of Unbounded & Bounded in Math?* SANTAMONICA, CA 90404, USA: Leaf in Math? Group, LTD., 2017, https://sciencing.com/meaningunbounded-bounded-math-8731294.html.

[41]   A. Francoeur, "Bounded regions," New York, NY 10038 USA, 2018, https://math.stackexchange.com/questions/253460/closed-sets-and-boundedsets.

[42]   C. Polanco, *Transformations: a mathematical approach - Fundamental concepts*. Sharjah, 7917, UAE: Bentham Science Publishers., 2017.

[43]   "Integrales dobles problemas resueltos," 13071 Ciudad Real, España, 2018, https://www.studocu.com/es/document/universidad-de-castilla-la-mancha/calculo-ii/ejercicios-obligatorios/problemas-resueltos-integrales-dobles/243326/view.

[44]   U. of Washington, "Department of mathematics," 2018, https://sites.math.washington.edu/ rothvoss/126D-spring-2014/lecture19-may12.pdf.

[45]   "Meaning of bounded and unbounded sets," Santa Monica, CA 90404, USA, 2018, https://sciencing.com/meaning-unbounded-bounded-math-8731294.html.

[46]   "Unbounded regions," New York, NY 10038 USA, 2018, https://math.stackexchange.com/questions/253460/closed-sets-and-boundedsets.

[47]   P. B. Laval, "Improper integrals," Kennesaw, GA 30144, USA, 2005, http://www.math.wisc.edu/ ~park/Fall2011/integration/Improper%20Integral.pdf.

[48] SlidePlayer, "Unbounded integrals," 2018, http://slideplayer.com/slide/8632436/.

[49] U. of Delaware, "Some examples of the use of Green's theorem," 2017, http://www.math.udel.edu/ angell/gren-exp.pdf.

[50] D. Q. Nykamp, "Green's theorem examples," 2018, https://mathinsight.org/greens theorem examples.

[51] U. del Páis Vasco, "Teorema de Stokes," 2018, http://www.ehu.eus/ mtpalezp/libros/07˙2.pdf.

[52] U. of Minnesota., "Divergence theorem examples," 2016, http://wwwusers. math.umn.edu/ nega0024/docs/2263 S14/GaussExamples.pdf.

[53] G. Wilkin, "Examples of Stokes' theorem and Gauss' divergence theorem," 2018, http://www.math.jhu.edu/ ˜graeme/files/math202 spring2009/StokesandGauss.pdf.

[54] R. Ablamowicz and G. Sobczyk, "Lecture series of clifford algebras and their applications," May 18 2002.

[55] D. Arapura, "Introduction to differential forms," 2016, https://www.math.purdue.edu/ dvb/preprints/diffforms.pdf.

[56] K. Bryan, "Differential forms," 2018, https://www.rosehulman.edu/ bryan/lottamath/difform.pdf.

[57] A. Castellanos-Moreno, *Introducción al Álgebra y al Cálculo Geométrico*. Departamento de Física, Universidad Autónoma de Sonora, 2013.

[58] U. A. de Madrid, "Álgebra tensorial – formas diferenciales," 2018, https://www.uam.es/otros/openmat/cursos/geodif/sections/geomivS14.pdf.

[59] H. J. A. Villamar, "Formas diferenciales," 2018, http://sistemas.fciencias.unam.mx/ erhc/c4/formas.pdf.

[60] S. Schmit and S. Gutzmacher, "Differential forms in Rn," 2015, https://www.mathi.uni-heidelberg.de/lee/Stehpan Sven.pdf.

[61] W. G. Faris, "Vector fields and differential forms," 2008, http://math.arizona.edu/ faris/mathanalweb/manifold.pdf.

[62] Wikipedia, "Maxwell-faraday equation," 2018, https://en.wikipedia.org/wiki/Faraday%27s law of induction.

SUBJECT INDEX